Math for Welders

by

Nino Marion
St. Clair College

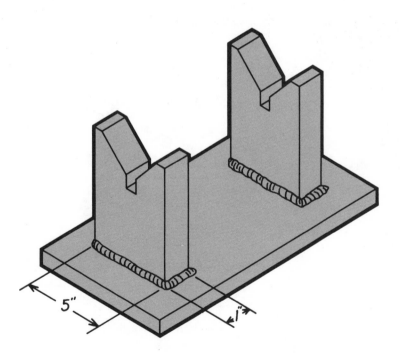

South Holland, Illinois
THE GOODHEART-WILLCOX COMPANY, INC.
Publishers

Library of Congress Catalog Card Number 89-28303
International Standard Book Number 0-87006-783-4

1234567890-89-9876543210

Library of Congress Cataloging in Publication Data

Marion, Nino.
  Math for welders / by Nino A. Marion.

    p. cm.
  ISBN 0-87006-783-4
  1. Welding—Mathematics. I. Title.
TS227.2.M37 1990
513'.08'8671—dc20   89-28303
                      CIP

INTRODUCTION

MATH FOR WELDERS is a combination textbook and workbook which teaches basic mathematics skills and provides practical exercises useful in the welding field. The textbook covers six areas of instruction including: Whole Numbers; Common Fractions; Decimal Fractions; Measurement; Percentages; and the Metric System. The topics are presented in a step-by-step approach with clear examples that makes learning easy and improves your understanding of the basics.

MATH FOR WELDERS lets you apply your learning using many drills and exercises. Space is provided in the book to work the problems and to record your answers. By referring to the answers to the odd numbered problems found in the back of the book, you will be able to check your progress as you study. The welding-related problems are designed to sharpen your application of basic mathematics to everyday situations.

An understanding of the material in this book is just as important to your career development as keeping your equipment in top shape or developing the ability to weld in various positions. Learn the skills taught here and become successful on the job!

Nino Marion

TABLE OF CONTENTS

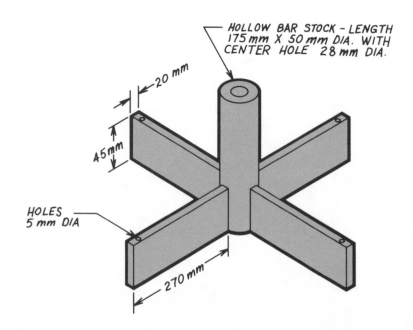

HOLLOW BAR STOCK - LENGTH 175 mm X 50 mm DIA. WITH CENTER HOLE 28 mm DIA.

20 mm

45 mm

HOLES 5 mm DIA

270 mm

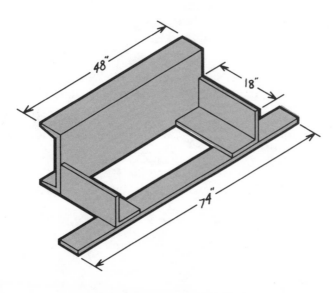

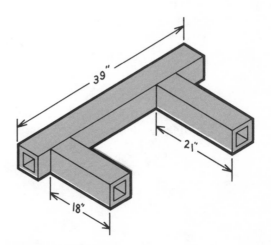

Section One

WHOLE NUMBERS

Objectives for Section One, Whole Numbers

After studying this section, you will be able to:

- Practice good math work habits
- Explain the Arabic number system
- Demonstrate how to round off whole numbers
- Define what is meant by a denominate number
- Perform addition of whole numbers
- Perform subtraction of whole numbers
- Perform multiplication of whole numbers
- Perform division of whole numbers
- Demonstrate how to check answers for four operation

Contents for Section One, Whole Numbers

INTRODUCTION TO WHOLE NUMBERS

INTRODUCTION

A mastery of mathematics is one of the skills expected of you as a welder. Fortunately, the method of acquiring math skills is no different than any other skill. You learn the principles and then practice them in repeated applications until they become "second nature."

What about the questions: "Do I need to know math if I know how to use an electronic calculator?" "Is it necessary to develop a skill in mental calculation and a mastery of math facts?" There are some practical reasons why the answer to these questions is YES. First, your employer, supervisor, etc., will expect you to be knowledgeable in math. They will assume that, as a trades-person, basic math skills and facts will be part of the mental skills you bring to the job. If you begin your training in math by relying on a calculator, you will do yourself a great disservice. A calculator will not give you the confidence and understanding that comes with building a skill.

Also, when a group of welders are discussing the math aspects of a job (dimensions, weights, volumes, costs, etc.) you must be able to "keep up" with the discussion. If your math skills are weak, you will likely feel very uncomfortable about joining in on this part of your work. Besides, there is certain admiration granted to a tradesperson by fellow workers who recognize that the person is clearly proficient in math.

MATH WORK HABITS

Although math is almost entirely a mental activity, there is a small but important tangible component to it. All your calculations, dimensions, and notes, must be readable. Correct results are, of course, the "final product" of math but the calculations should be done with a high degree of neatness and care. Orderly work in math is one of the important factors in arriving at correct answers. Listed are some suggestions you might find useful.

1. Write your numbers fairly large.
2. A poorly written 4 and 9 are easily confused.
3. Try writing 7 as 7. The number 7, when handwritten, can sometimes be confused with 1 or 2. The slash helps to avoid this confusion.
4. Always carry a thick, sharp pencil while in the shop or in the field.
5. When dealing with numbers, don't rush. In fact, deliberately slow yourself down. If you consider the amount of time it takes to turn out a job in the shop, there is no measurable gain in time by calculating and writing figures quickly.

OUR NUMBER SYSTEM

Our number system is based on ten digits, namely, 0, 1, 2, 3, 4, 5, 6, 7, 8, 9. Because the Arabs originated the system, it is sometimes called the **Arabic number system.** Also, since it consists of ten digits, it is often referred to as the **Decimal number system.** Decimal is a word derived from Latin meaning "ten."

Practically any number can be expressed by arranging these numbers in a certain order. When a series of digits is written, such as 2497, each of the numbers in that group take on a certain value based on its **place value** in the lineup. In the example 2497, the 7 is considered a 7 because it is in the ones (or units) position. The 9 has a value of 90 because it is in the tens position. You can think of it as $9 \times 10 = 90$. The 4, because of its place in the lineup, has a value of

400. It is in the hundreds column, so you can think of it as $4 \times 100 = 400$. The 2 is placed in the thousands position and has a value of 2000.

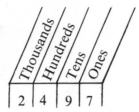

Each of the positions in a line of figures has a value and a name. Listed below are some of the names and their place position.

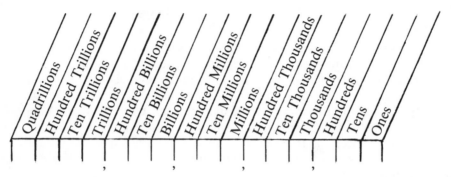

To help make numbers easier to read, a comma is usually placed after every third digit counting from the right. Here is an example. The number 3,894,076,215 is divided by commas and is read as three billion, eight hundred ninety four million, seventy six thousand, two hundred fifteen. The 3 has a place value of three billion. The 8 has a place value of eight hundred million, the 9 has a place value of ninety million and so on.

ROUNDING NUMBERS

There are some occasions in math where extreme accuracy is not required. Cost estimates for most jobs are not usually exact. Precise weights of large weldments are often not necessary. In these and many other situations, figures may be **rounded.** Since rounded figures are not perfectly accurate, they are called **approximate numbers.**

HOW TO ROUND NUMBERS

Here is method for rounding numbers:
1. First, find out to what place value you are rounding. Normally, you are asked to "round to the nearest hundred or thousand," etc.
2. Place a little tick mark over the digit in that place position.
3. If the figure to the right of that number is 5 or more, increase the ticked number by one.
4. If the figure to the right is less than five, do not change the ticked number.
   For example: Round 17,289,364 to the nearest ten thousand. Place a tick mark at the ten thousand position (17,289,364). Since the digit to the right is "5 or more," the 8 is changed to 9. The rounded answer is 17,290,000.

DENOMINATE NUMBERS

Despite its long name, denominate numbers are simple. They are just numbers that represent measurements. For example, 4 feet is a denominate number. The number 4, by itself, is not a denominate number. As soon as you add a description to a number showing that it represents some kind of measurement, then it is classified a **denominate number.** More examples are, 6 gallons, 5 3/16 in., 186 pounds, 1,800°F, and 55 miles per hour.

As you may have guessed, welders work almost entirely with denominate numbers. It is important to remember that when a question is expressed in denominate numbers, your answer must also be a denominate number.

KEY TERMS FOR WELDERS
ARABIC NUMBER SYSTEM
DECIMAL NUMBER SYSTEM
PLACE VALUE
ROUNDING
APPROXIMATE NUMBERS
DENOMINATE NUMBERS

Unit 1—Practicing Using Whole Numbers

1. What are the place values of each figure in the following numbers?

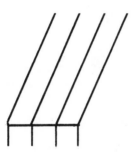

(a) 471

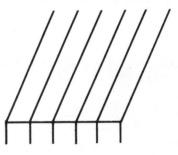

(b) 15,286

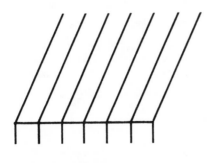

(c) 349,015

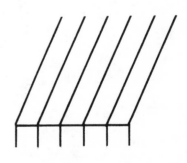

(d) 27,636

2. Round the following numbers to the nearest hundred:
   (a) 694,701   (b) 849

   (c) 17,213   (d) 9,999

   (e) 29,897   (f) 4,509

3. Round the following numbers to the nearest thousand:
   (a) 11,195   (b) 449,561

   (c) 449,156   (d) 85,028

   (e) 1,294,637   (f) 9,195

4. Which of the following are denominate numbers?
   (a) 250 PSI   (b) 11,500 sq. ft.

   (c) five acres   (d) 17.5 tons

   (e) 1,895   (f) 0.525

   (g) $7.65   (h) $7.65 per hour

   (i) 40 inches per minute   (j) E-6015

   (k) 12   (l) 12 dozen

ADDITION OF WHOLE NUMBERS

INTRODUCTION

Whole numbers are manipulated in arithmetic by four **basic operations:** addition, subtraction, multiplication, and division. The most widely used of these four operations is addition. You will learn about addition in this unit. The other three basic operations will be studied in following units.

METHOD USED TO ADD WHOLE NUMBERS

Whole numbers are added by stacking the numbers to be added in a column with all the units on the right side lined up one upon the other.

```
   units              units              units
     ↓                  ↓                  ↓
  4,956                 27                354
     21                192              2,032
    191                 11                111
    256              1,542              9,461
 10,431                  5                 17
```

It is very, very important that all of the digits for each place value line up exactly above other digits of the same place value. Refer back to the illustration in the first unit on place values.

Adding the units column in the example below results in 15. The 5 is placed under the ones column and the 1 is carried to the next column. When you do this, write it at the top of the tens column and make it smaller than your other digits. Next, add the tens column. The result is 26. Write in the number 6, and carry the 2 to the hundreds column. Continue in this manner until all columns are added. The final answer is called the **sum.** Always check your accuracy.

```
   2 2 1
  9,966
     21
    991
    256
 94,531
105,765
```

CHECKING ADDITION BY ADDING UP AND DOWN

Always check your math work. If you added by going up the column, then check it by adding going down.

LOOK FOR TENS

When adding a column, search for combinations of numbers that add up to 10. In the column below, 8 plus 2 equals 10, and 1 plus 9 equals 10 (that's 20, so far), and 4 plus 6 equals 10 (totaling

30). The remaining digit is 7, so the total is 37. You will find it easier doing the mental calculations this way. Be sure to tick off the numbers as you add them or you may forget which ones you have already added:

$$
\begin{array}{r}
9 \\
7 \\
2\checkmark \\
6 \\
4 \\
1 \\
\underline{8\checkmark} \\
37
\end{array}
$$

MARK THE ANSWER

When you arrive at a final answer, mark it in a very clear manner. Try boxing it in, like this
248 or double underlining it 248. This is especially important if you've got a mass of numbers on a sheet mixed in with previous, unrelated calcuations. This is sometimes the situation when doing calculations in the shop or in the field.

DENOMINATE NUMBERS

Be alert for denominate numbers. If a question is expressed in inches, your answer should also be in inches.

KEY TERMS FOR WELDERS
BASIC OPERATIONS SUM

Unit 2—Practicing Addition of Whole Numbers

Show all your work. Be certain the columns line up. Box your answers.

1. Add the following:

(a)	235 471	(b)	6,241 7,356	(c)	964 35,309

(d)	102 628 754 888 973	(e)	75 5,491 67,811 604 15,473	(f)	11,009 7,489,621 18,219,305 47,951 1,987

	(g)	8	(h)	46	(i)	167,540
		9		325		22,814
		7		3		499,863
		6		1,588		200
		2		231		98,157
		4		91		5
		1		942		37,254

2. Add the following:

   (a) 7 + 9 + 1 + 3 + 3 + 8 + 5 =

   (b) 904 + 214 + 22 =

   (c) 18,961 + 718 + 6,800 =

   (d) 6,525 + 7,182 + 293 =

   (e) 966 + 372 + 165 + 638 + 300 + 200 =

   (f) 731 + 82 + 234 + 2,699 + 523 + 64 =

   (g) 18 + 444 + 27,981 + 1,234 + 75,211 + 7 =

   (h) 21,987 + 101,001 + 622 + 9 + 36,299 + 981,202 =

   (i) 4 + 144 + 414,411 + 69,835 + 641,538 + 99 =

3. Three pieces of double extra-strong pipe in inventory have the following lengths: 27 in., 42 in., and 19 in. What is the total length of double extra-strong pipe in inventory?

4. Three welded frames are to shipped by truck to a customer. The smallest frame weighs 1,895 lbs., the next frame weighs 2,982 lbs., and the largest weighs 3,206 lbs. What is the total weight?

5. A manufacturer of welding rods produced the following quantities: January, 189,261; February, 72,450; March, 284,361. What was the total production for the three months?

6. World coal production for each of the past six years has been 8,523,985,000 tons, 8,405,470,000 tons, 8,311,903,000 tons, 8,067,878,000 tons, 7,646,980,000 tons, and 6,996,037,000 tons. What was the world production of coal for the past six years?

7. Alexander Lincoln worked the number of hours shown below for the month of December. What is the total number of hours he worked for the month?

DECEMBER								
WEEK	DATE	HOURS	WEEK	DATE	HOURS	WEEK	DATE	HOURS
1	1	8	2	10	8	4	21	7
1	2	8	2	11	6	4	22	8
1	3	8				4	23	6
1	4	8	3	14	4	4	24	3
1	5	4	3	15	8			
			3	16	8	5	28	5
2	7	6	3	17	8	5	29	6
2	8	8	3	18	7	5	30	9
2	9	8	3	19	5	5	31	3

8. A large factory has 2,816 men and 1,042 women in the Fabricating Department; 110 men and 119 women in the Repair and Maintenance Department; 39 men and 63 women in the Administration Department; 367 men and 457 women in the Engineering Department; 98 men and 85 women in the Sales and Service Department. How many men are employed by the company? How many women? What is the total number of people employed?

9. What is the total length of angle iron?

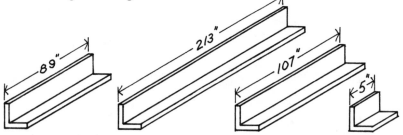

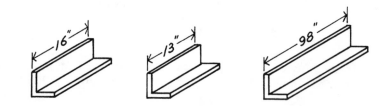

10. What is the total length of the notched plate?

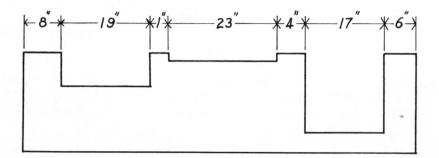

11. What is the total distance between the centers of the holes at each end of the part?

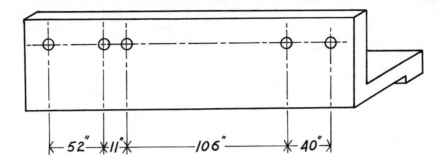

12. What is the total length of square tubing required for the weldment?

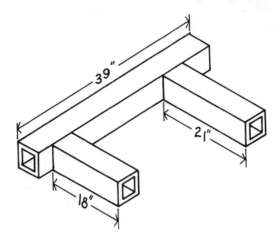

13. This pie chart illustrates the number of hours required to complete various phases of Job Order #8805. What is the total number of hours required to complete the job?

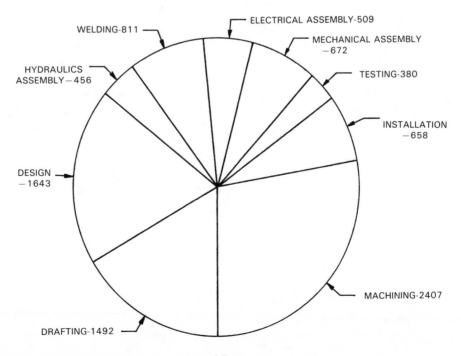

14. What is the total steel production for the following countries?

COUNTRY	PRODUCTION IN TONS
U.S.A.	67,746,000
CANADA	14,825,000
CHINA	35,620,000
CZECHOSLOVAKIA	15,319,000
FRANCE	21,346,000
W.GERMANY	41,662,000
ITALY	25,085,000
JAPAN	101,708,000
POLAND	15,112,000
ROMANIA	13,193,000
U.S.S.R.	148,980,000
U.K.	15,667,000

Unit 3

SUBTRACTION OF WHOLE NUMBERS

INTRODUCTION

Subtraction is one of the four basic math operations. **Subtraction** is the process of finding the difference between two numbers.

METHOD USED TO SUBTRACT WHOLE NUMBERS

Two whole numbers are subtracted by stacking them one upon the other with the units columns lined up on the right side. The larger number belongs on the top position. Begin subtracting at the units column and work your way through the problem, column by column. Here are some examples:

$$\begin{array}{r} 84 \\ -23 \\ \hline \end{array} \qquad \begin{array}{r} 5{,}794 \\ -\ 62 \\ \hline \end{array}$$

As with most math operations, there are new words to learn. Below are names of the parts of a subtraction operation:

$$\begin{array}{r} 1{,}347 \\ -\ 26 \\ \hline 1{,}321 \end{array} \begin{array}{l} \leftarrow \textbf{Minuend} \\ \leftarrow \textbf{Subtrahend} \\ \leftarrow \textbf{Difference or Remainder} \end{array}$$

You will often encounter a situation like this:

$$\begin{array}{r} 47 \\ -19 \\ \hline \end{array}$$

The units column displays 7 minus 9. Since 9 cannot be subtracted from 7, you must **borrow** a number from the place value to the left. Here's how. Borrow 1 from the 4 in the tens column, draw a line through the 4, and replace it with 3. Now, place a tiny 1 next to the 7 to create the number 17:

$$\begin{array}{r} ^{3}\cancel{4}7 \\ -19 \\ \hline \end{array}$$

Now, subtract 9 from 17 to get 8. Then, to finish the problem, subtract 1 from 3 to get 2:

$$\begin{array}{r} ^{3}\cancel{4}7 \\ -19 \\ \hline 28 \end{array}$$

Here is another more complicated example:

$$1,007$$
$$-99$$

In the units column, you cannot subtract 9 from 7. As in the previous example, you look to the tens column to borrow a 1 and find there are no numbers to borrow. You then have to go to the hundreds column and still there is no number to borrow. Finally, in the thousands column, you can borrow a 1. By borrowing 1, the existing 1 becomes 0. Draw a line through the existing 1 (it becomes 0). Now, place a tiny 1 next to the 0 in the hundreds column to create the number 10. Then borrow 1 from it and reduce it to 9 by crossing out the 10 and replacing it with a 9. Now, move to the tens column with the borrowed 1. Place a tiny 1 next to the 0 in the tens column to create the number 10. Borrow 1 making it 9 and transfer the 1 to the units column, thereby changing the 7 to 17. You can now begin the problem by subtracting 9 from 17 in the units column:

$$1,007$$
$$-\ \ 99$$
$$908$$

When speaking or writing about subtraction, there are a number of ways of expressing yourself. If you are not familiar with them, it could be difficult to figure out just what is being subtracted from what! Here are examples of how you are likely to see a subtraction expressed:

24 take away 11
24 less 11
24 minus 11
24 subtract 11
24 reduced by 11
Find the difference between 24 and 11

Always check your accuracy.

CHECKING SUBTRACTION BY ADDING

Check your work by adding the difference and the subtrahend. The total should equal the minuend:

	198,462
Subtrahend	− 76,291
+ Difference	122,171
= Minuend	198,462

KEY TERMS FOR WELDERS
MINUEND
SUBTRAHEND
DIFFERENCE
REMAINDER
BORROW

Unit 3—Practicing Subtraction of Whole Numbers

Show all your work. Be certain the columns line up. Box your answers.

1. Subtract the following:

(a)   875
         -211

(b)   1,280
         $-1,120$

(c)   9,706
         $-7,960$

(d)   74,240
         -659

(e)   21,008
         -989

(f)   12,345
         $-6,789$

(g)   45,000
         $-2,370$

(h)   894,216
         $-735,094$

(i)   1,019,471
         $-4,048$

2. Subtract the following:

(a) $847 - 305 =$

(b) 6,544 minus 822 =

(c) Take 28,001 away from 29,477 =

(d) Find the difference between 18,234 and 11,012 =

(e) Find the result when 707 is taken away from 747 =

(f) What is the difference between 74 and 47 =

(g) Subtract 7,641 from 22,350 =

(h) 184,555 less 90,915 =

(i) Reduce 19,486 by 215 =

3. A shop had 2,024 brackets in stock on February 15th. One month later, there were 820 remaining in inventory. How many brackets were used during the month?

4. A fabricating shop had the following steel in inventory:

> 15,285 lbs. of 22 gage steel
> 37,549 lbs. of 16 gage steel
> 89,041 lbs. of 14 gage steel

One large customer order used the following quantities:

> 6,102 lbs. of 14 gage steel
> 11,910 lbs. of 22 gage steel
> 2,819 lbs. of 16 gage steel

(a) What weight of steel of each classification remained after the order was completed?

(b) What was the total weight of the customer order?

5. Proto Mfg. produced the following welding tips in one week:

Tip Size	Quantity Produced
#68	118,295
#51	7,050
#35	892

The following quantities were shipped to various welding supply houses:

Tip Size	Quantity Shipped
#35	0
#68	73,460
#51	5,070

How many of each tip size remain at Proto?

6. The original design for a large machine calls for a frame of structural steel tubing that weighs 4,851 lbs. An engineering change was later approved, adding additional stiffeners. The final weight of the frame was 5,212 lbs. What is the weight of the additional stiffeners?

7. In 1983, there were 225,185 service stations in the U.S. By 1987, this number had declined by 47,242. Among the stations in business in 1987, there were 58,361 that did not use welding equipment. How many service stations in 1987 used welding equipment?

8. A coal car was redesigned by replacing some of the steel with aluminum. The designers wanted to reduce its weight and also increase its payload to 110 tons. The total weight of the redesigned car was 42,695 pounds. The amount of aluminum used was 9,860 pounds and the remainder was steel. What weight of steel was used?

9. Three pieces were cut from a length of cold-rolled steel. What length remains? Ignore loss of material due to cutting.

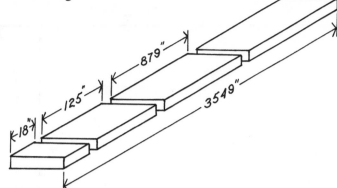

10. What is the distance between the centers of the holes?

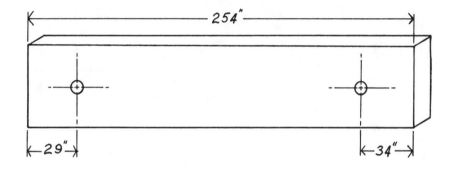

11. What is the height of the vertical piece (A) ?

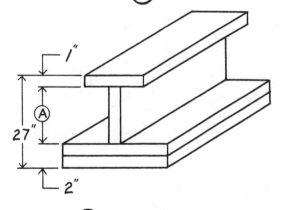

12. What is the length of distance (A) ?

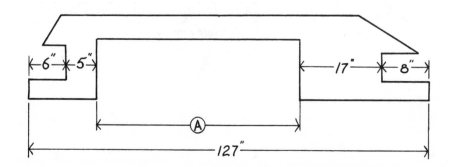

13. What is the difference 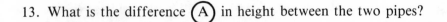 in height between the two pipes?

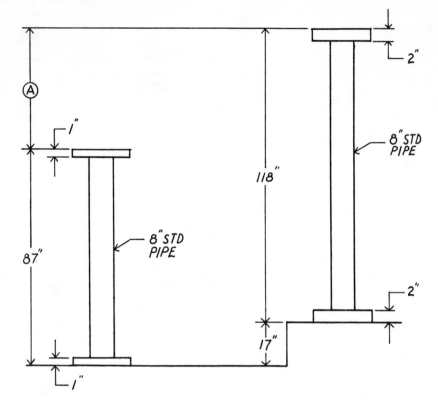

14. Prepare a list of the inventory remaining after Job #8890 is completed.

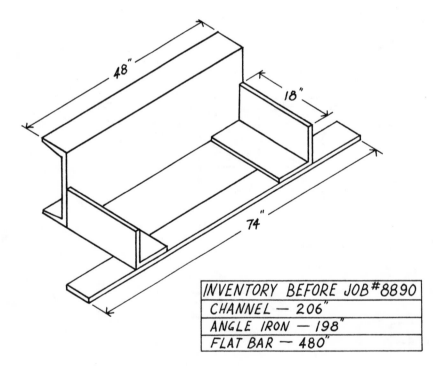

INVENTORY BEFORE JOB #8890		
CHANNEL — 206"		
ANGLE IRON — 198"		
FLAT BAR — 480"		

MULTIPLICATION OF WHOLE NUMBERS

INTRODUCTION

Multiplication can be thought of as a fast way to do addition. That may seem odd at first, but look at it this way. If you were to add 3 + 3 + 3 + 3 + 3, the total, of course, would be 15. However, you could think of it as, "3 taken 5 times" or "5 times the number 3." This way of thinking works well for mental calculations of small numbers. It speeds up the process and simplifies the problem. Many people have memorized the multiplication facts for numbers 1 through 12 and these are often seen in chart form as multiplication tables.

MULTIPLICATION TABLE

2x1 = 2	3x1 = 3	4x1 = 4	5x1 = 5	6x1 = 6	7x1 = 7	
2x2 = 4	3x2 = 6	4x2 = 8	5x2 = 10	6x2 = 12	7x2 = 14	
2x3 = 6	3x3 = 9	4x3 = 12	5x3 = 15	6x3 = 18	7x3 = 21	
2x4 = 8	3x4 = 12	4x4 = 16	5x4 = 20	6x4 = 24	7x4 = 28	
2x5 = 10	3x5 = 15	4x5 = 20	5x5 = 25	6x5 = 30	7x5 = 35	
2x6 = 12	3x6 = 18	4x6 = 24	5x6 = 30	6x6 = 36	7x6 = 42	
2x7 = 14	3x7 = 21	4x7 = 28	5x7 = 35	6x7 = 42	7x7 = 49	
2x8 = 16	3x8 = 24	4x8 = 32	5x8 = 40	6x8 = 48	7x8 = 56	
2x9 = 18	3x9 = 27	4x9 = 36	5x9 = 45	6x9 = 54	7x9 = 63	
2x10 = 20	3x10 = 30	4x10 = 40	5x10 = 50	6x10 = 60	7x10 = 70	
2x11 = 22	3x11 = 33	4x11 = 44	5x11 = 55	6x11 = 66	7x11 = 77	
2x12 = 24	3x12 = 36	4x12 = 48	5x12 = 60	6x12 = 72	7x12 = 84	
8x1 = 8	9x1 = 9	10x1 = 10	11x1 = 11	12x1 = 12		
8x2 = 16	9x2 = 18	10x2 = 20	11x2 = 22	12x2 = 24		
8x3 = 24	9x3 = 27	10x3 = 30	11x3 = 33	12x3 = 36		
8x4 = 32	9x4 = 36	10x4 = 40	11x4 = 44	12x4 = 48		
8x5 = 40	9x5 = 45	10x5 = 50	11x5 = 55	12x5 = 60		
8x6 = 48	9x6 = 54	10x6 = 60	11x6 = 66	12x6 = 72		
8x7 = 56	9x7 = 63	10x7 = 70	11x7 = 77	12x7 = 84		
8x8 = 64	9x8 = 72	10x8 = 80	11x8 = 88	12x8 = 96		
8x9 = 72	9x9 = 81	10x9 = 90	11x9 = 99	12x9 = 108		
8x10 = 80	9x10 = 90	10x10 = 100	11x10 = 110	12x10 = 120		
8x11 = 88	9x11 = 99	10x11 = 110	11x11 = 121	12x11 = 132		
8x12 = 96	9x12 = 108	10x12 = 120	11x12 = 132	12x12 = 144		

1	2	3	4	5	6	7	8	9	10	11	12
2	4	6	8	10	12	14	16	18	20	22	24
3	6	9	12	15	18	21	24	27	30	33	36
4	8	12	16	20	24	28	32	36	40	44	48
5	10	15	20	25	30	35	40	45	50	55	60
6	12	18	24	30	36	42	48	54	60	66	72
7	14	21	28	35	42	49	56	63	70	77	84
8	16	24	32	40	48	56	64	72	80	88	96
9	18	27	36	45	54	63	72	81	90	99	108
10	20	30	40	50	60	70	80	90	100	110	120
11	22	33	44	55	66	77	88	99	110	121	132
12	24	36	48	60	72	84	96	108	120	132	144

To use this chart, the number on the top row multiplied by the number on the left column results in the product found where the row and column intersect.

Incidentally, there is no special reason why the multiplication table stopped at 12. It is probably because most people did not want to memorize beyond that point. Even in this age of electronic calculators, you should be very familiar with the multiplication table. You should commit it to memory.

METHOD USED TO MULTIPLY WHOLE NUMBERS

Large numbers are calculated by writing them out. To begin, you should know the appropriate terminology.

$$132 \leftarrow \textbf{Multiplicand}$$
$$\times 23 \leftarrow \textbf{Multiplier}$$

Refer to the multiplication tables if you need to refresh your skills as you study these examples.

Always line up the figures on the right side in the units column. In this typical example, begin by multiplying every digit in the multiplicand by the 3 in the multiplier.

$$
\begin{array}{r}
132 \\
\times 23 \\
\hline
396
\end{array}
$$

Now, multiply every digit in the multiplicand by the 2 in the multiplier, indenting the answer so it lines up directly under the 2 in the multiplier. Then, add the results.

$$
\begin{array}{r}
132 \\
\times\ 23 \\
\hline
396 \\
264\ \\
\hline
3036
\end{array}
$$

Normally, you will have to **carry** numbers, as in the next example.

$$298 \times 27$$

In the first operation, $7 \times 8 = 56$. Place the 6 in the answer and write a little 5 directly above the 9. Now, multiply 7×9 to get 63 and add (carry) the 5 to it, arriving at 68. Place 8 in the answer and a little 6 above the 2. Multiply 7×2 to get 14. Add the carried 6 to the 14 and write 20 in the answer. Always check your accuracy.

$$
\begin{array}{r}
\overset{6\,5}{298} \\
\times\ 27 \\
\hline
2086 \\
596 \\
\hline
8046
\end{array}
$$

CHECKING MULTIPLICATION BY REVERSE POSITIONS

A method of checking your work is to reverse the positions of the multiplier and multiplicand and repeat the problem.

ACCURATE ALIGNMENT

Be sure to accurately line up the numbers in your answer. You may include right-hand zeros to help keep the figures aligned.

$$
\begin{array}{r}
154 \\
\times\ 368 \\
\hline
1232 \\
9240 \\
46200 \\
\hline
56672
\end{array}
$$

SMALLER MULTIPLIER

It does not matter which number you position as the multiplier or multiplicand.

$$
\begin{array}{r}
1286 \\
\times 27 \\
\end{array}
\qquad\qquad
\begin{array}{r}
27 \\
\times 1286 \\
\end{array}
$$

Both will provide the same results. You will find it easier, however, if you place the smaller number as the multiplier.

KEY TERMS FOR WELDERS
MULTIPLICAND
MULTIPLIER
CARRY

Unit 4—Practicing Multiplication of Whole Numbers

Show all your work. Be certain the columns line up. Box your answers.

1. Multiply the following:

   (a)
   $$\begin{array}{r} 12 \\ \times\ 34 \\ \hline \end{array}$$

   (b)
   $$\begin{array}{r} 98 \\ \times\ 75 \\ \hline \end{array}$$

   (c)
   $$\begin{array}{r} 56 \\ \times\ 70 \\ \hline \end{array}$$

   (d)
   $$\begin{array}{r} 659 \\ \times\ 423 \\ \hline \end{array}$$

   (e)
   $$\begin{array}{r} 207 \\ \times\ 845 \\ \hline \end{array}$$

   (f)
   $$\begin{array}{r} 913 \\ \times\ 600 \\ \hline \end{array}$$

   (g)
   $$\begin{array}{r} 84,796 \\ \times\ 56 \\ \hline \end{array}$$

   (h)
   $$\begin{array}{r} 93,253 \\ \times\ 24 \\ \hline \end{array}$$

   (i)
   $$\begin{array}{r} 53,602 \\ \times\ 35 \\ \hline \end{array}$$

2. Multiply the following:

   (a) $523 \times 9,642 =$

   (b) $804 \times 5,022 =$

   (c) $68 \times 12,673 =$

   (d) $67,691 \times 37 =$

   (e) $12 \times 24 \times 36 =$

   (f) $18 \times 13 \times 20 =$

(g) $109 \times 14 \times 34 =$

(h) $263 \times 101 \times 24 =$

(i) $468 \times 219 \times 153 =$

3. Two hundred and thirty-nine base frames were produced for a special order. Each frame required ninety-eight spot welds. What is the total number of spot welds for the entire job?

4. How many rivets are in 3 cartons if each carton contains 144 boxes and each box contains 48 rivets?

5. Thirty-nine rows of studs will be welded to a metal platform. Each row will be 5 in. apart. There will be twenty-seven studs in each row. What is the total number of studs required?

6. A GMAW welder traveling at 19 inches per minute was in constant use for 5 1/2 hours each day for 24 days. How many inches of weld were deposited in that time?

7. The roof of an arena built for the University of Iowa required 5,237 plates to secure the roof trusses to the support beams. The plates were flame cut to shape, then subcontracted to a machine shop for drilling the following bolt holes:

> 1,728 plates with 6 holes
> 295 plates with 5 holes
> 964 plates with 9 holes
> 2,235 plates with 14 holes
> 15 plates with 23 holes

What is the total number of bolt holes drilled?

8. A power line built for Louisiana Light and Power Co. had 619 towers constructed of various sizes of pipe. Specifications for the job are listed below.
   (a) Find the total length of pipe for each size.

   (b) Find the total length of pipe for the entire project.

STEEL PIPE REQUIREMENTS			
	TOWER LEGS	**CROSS ARMS**	**STIFFENERS**
MATERIAL SIZE NUMBER PER TOWER LENGTH	ASTM A595 2′ DIAMETER 2 175 TOWERS AT 93′ 300 TOWERS AT 108′ 144 TOWERS AT 122′	ASTM A595 1 1/2′ DIAMETER 1 102′	ASTM A595 1′ DIAMETER 2 35′

9. These equally sized rings are welded together.
   (a) What is the total length of the weldment?

   (b) What is the total weight of the weldment if each ring weighs 6 pounds?

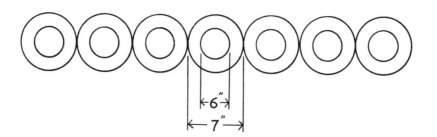

10. How many of the following GMAW welding tips can be produced in 21 working days at 9 hours per day?

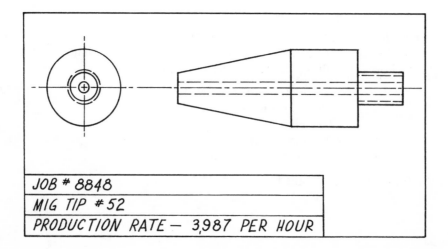

JOB # 8848
MIG TIP # 52
PRODUCTION RATE — 3,987 PER HOUR

11. How many inches of chrome plated tubing will be required to manufacture 289 tables?

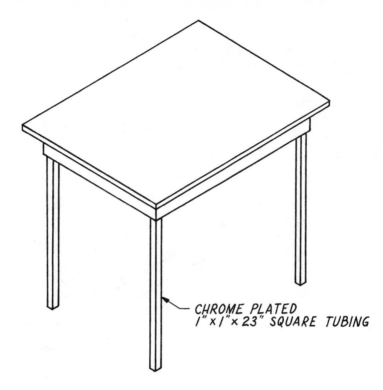

CHROME PLATED
1" x 1" x 23" SQUARE TUBING

12. Each of the posts in this shaft support are welded all around. How many inches of weld will be applied to make 2,893 shaft supports?

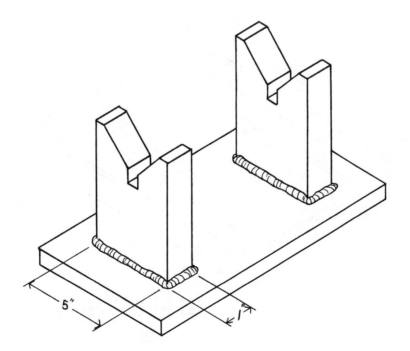

5"

1"

13. Calculate the weight of each of the following weldments. Assume there is no filler added to the welds.

(a)

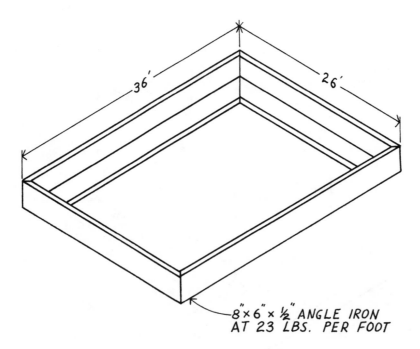

2½" × 2½" × 5/16" ANGLE IRON AT 5 LBS. PER FOOT

8" × 8" × 7/8" ANGLE IRON AT 45 LBS. PER FOOT

15'

18'

13. (b)

36'

26'

8" × 6" × ½" ANGLE IRON AT 23 LBS. PER FOOT

13. (c)

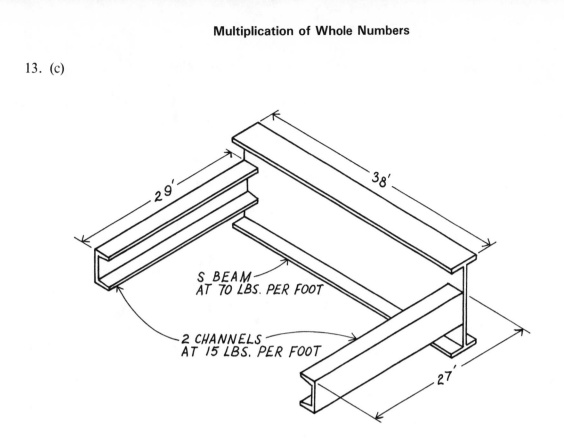

14. Eighty-five frames are to be fabricated. What is the total number of inches of tubing required.

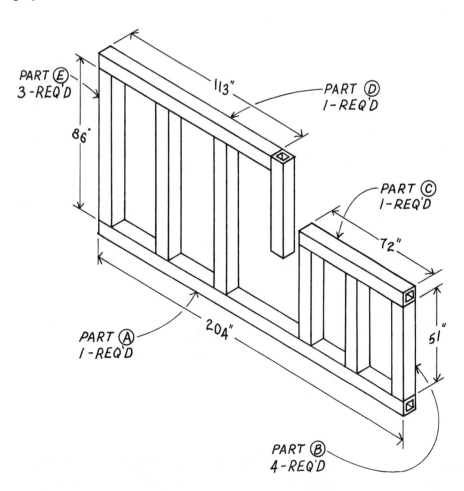

DIVISION OF WHOLE NUMBERS

INTRODUCTION

Division is the process of finding how many times one number can be "contained" in another number. The number 50, for example, "contains" the number 10, five times. There are a variety of expressions used to describe this relationship.

> There are 5 tens in 50.
> Ten "goes into" 50, five times.
> Fifty divided by 10 equals five.
> Ten divides into 50, five times.

While there are a number of ways of expressing this in spoken English, there are also a number of ways of symbolizing division. All of the following symbolize the dividing of 1188 by 9.

$$9\overline{)1188}$$

$$1188 \div 9$$

$$\frac{1188}{9}$$

$$1188/9$$

METHOD USED TO DIVIDE WHOLE NUMBERS

To begin, you should know the appropriate terminology.

$$\text{Divisor} \rightarrow 9\overline{)1188} \begin{matrix} \leftarrow \textbf{Quotient} \\ \leftarrow \textbf{Dividend} \end{matrix}$$

with 132 above as the Quotient.

These words (like many math terms) may seem awkward at first. You will see them mainly in math books and occasionally in some formula tables. They are needed to help provide a smooth explanation of the division process.

To perform a division operation, cover all of the dividend, except the first digit, with your finger.

$$7\overline{)131138}$$

Then, ask yourself, "will the divisor go into this digit?" If yes, then ask how many times. In this example, the answer is, of course, "no." Now, uncover the next digit and ask if 7 will divide into 13 and how often. The answer is yes, once. Write the 1 above the 3 as shown. It is extremely important to line up the numbers accurately.

$$
\begin{array}{r}
1 \\
7\,\overline{)131138}
\end{array}
$$

The next step is to multiply 7×1 and write the answer under the 13.

$$
\begin{array}{r}
1 \\
7\,\overline{)131138} \\
7
\end{array}
$$

Next step, subtract 7 from 13.

$$
\begin{array}{r}
1 \\
7\,\overline{)131138} \\
\underline{7} \\
6
\end{array}
$$

"Borrow" or "bring down" the next digit in the dividend. Place a small tick mark under it as a reminder that you have already picked it up and placed it next to the 6 to form the number 61.

$$
\begin{array}{r}
1 \\
7\,\overline{)131138} \\
7 \\
61
\end{array}
$$

Again, "Does the divisor go into 61?" The answer is yes, eight times.

$$
\begin{array}{r}
18 \\
7\,\overline{)131138} \\
7 \\
61 \\
56
\end{array}
$$

Be sure the 8 is lined up directly above the 1 in the dividend. Multiply 7×8 to get 56. Subtract 56 from 61 to get 5.

$$
\begin{array}{r}
18 \\
7\,\overline{)131138} \\
7 \\
61 \\
\underline{56} \\
5
\end{array}
$$

Continue repeating this cycle with each of the remaining digits.

$$
\begin{array}{r}
18734 \\
7\overline{)131138} \\
7 \\
\hline
61 \\
56 \\
\hline
51 \\
49 \\
\hline
23 \\
21 \\
\hline
28 \\
28 \\
\hline
0
\end{array}
$$

That is all there is to it, except for two variations you will encounter. First, there will be some occasions where the divisor cannot divide into the number that has been "brought down."

$$
\begin{array}{r}
1 \\
7\overline{)728} \\
7 \\
\hline
02
\end{array}
$$

The number 2 has already been "brought down," but will the divisor go into 2? The answer is "no." In this case, place a zero above the 2 and bring down the next number.

$$
\begin{array}{r}
10 \\
7\overline{)728} \\
7 \\
\hline
028
\end{array}
$$

Forgetting to place this zero in the quotient is an error that a beginner can easily make. Now, proceed with the problem.

$$
\begin{array}{r}
104 \\
7\overline{)728} \\
7 \\
\hline
028 \\
28 \\
\hline
0
\end{array}
$$

Second, there will be many occasions where the divisor does not divide evenly into the dividend. In these cases, you will have a **remainder.**

$$
\begin{array}{r}
259 \\
5\overline{)1296} \\
10 \\
\hline
29 \\
25 \\
\hline
46 \\
45 \\
\hline
1 \text{ R}
\end{array}
$$

The remainder here is 1. You might find it useful to print a tiny R next to the remainder. Always check your accuracy.

CHECKING DIVISION BY MULTIPLYING

Check your work by multiplying the quotient by the divisor. If there was a remainder, add it to the result and the answer should be the same as the dividend.

$$
\begin{array}{r}
259 \\
\times\ 5 \\
\hline
1295 \\
+\ 1 \\
\hline
1296
\end{array}
$$

DO NOT CROWD

As you can see, division problems tend to use a lot of space. Always give yourself plenty of room. Do not try to squeeze the figures into a small space, this will only increase your chances of making an error.

KEY TERMS FOR WELDERS
DIVISOR
DIVIDEND
QUOTIENT
REMAINDER

Unit 5—Practicing Division of Whole Numbers

1. Divide the following:

(a) $8\overline{)216}$ (b) $29\overline{)348}$ (c) $6\overline{)618}$

(d) $7\overline{)8,050}$ (e) $9\overline{)81,064}$ (f) $35\overline{)2,410}$

(g) $143\overline{)1,001}$ (h) $365\overline{)44,895}$ (i) $400\overline{)160,000}$

2. Divide the following:
   (a) 4,826 ÷ 3,519 =

   (b) 1,840 divided by 72 =

   (c) 1,066$\overline{)981,734}$

   (d) 18,604/290 =

   (e) 1,776$\overline{)3,528,912}$

   (f) 245 divided into 6,195 =

   (g) 8,521$\overline{)340,915,264}$

   (h) How many 335's are there in 6,369?

   (i) Divide 648 into 8,759

3. A company which manufactures electric light fixtures used 11,385 lbs. of solder in one year. If there were 253 working days in the year, how many pounds of solder would be consumed on an average day?

4. Welding a heavy steel plate with a V-groove joint required 1,683 ounces of filler. The plate was 3 in. thick and 187 in. long. How many ounces of filler were used for each inch of weld?

5. A piece of angle iron 177 in. long is to be cut into 11 in. pieces. Assuming there is no loss of material due to cutting, how many pieces would be obtained?

6. A fabricating shop was successful in their bid for the contract to supply railings for a bridge across the Detroit River. The bridge was to be 5,346 ft. long and railings were to be provided for both sides of the bridge. The railings were welded in 18 ft. sections and delivered to the job site.
   (a) How many sections of railing were built?
   (b) If each section weighed 642 lbs., what was the total weight of the railings?

7. A supply house ordered 14 cartons of aluminum flux in cans. Each carton weighed 216 lbs. (excluding the carton). If each can weighed 3 lbs., how many cans of flux are in the order?

8. During a recent repair of the Golden Gate bridge, 29,040 feet of on-site welding was laid. The job was completed in 169 days. What was the average number of feet of weld laid per day? Express your answer to the nearest foot.

9. What is the distance between the centers of the evenly spaced holes shown below?

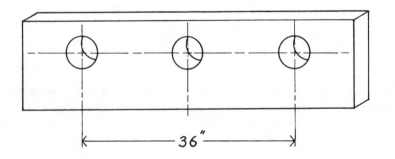

10. What is the distance between the centers of the evenly spaced holes?

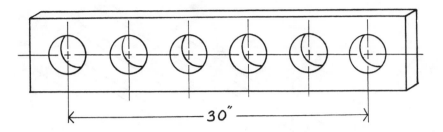

11. Find the distance between the holes.

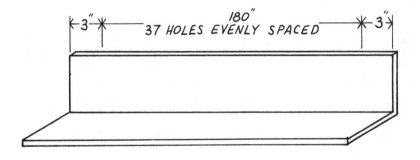

12. How many 17 in. pieces can be cut from this 16,761 in. roll of steel?

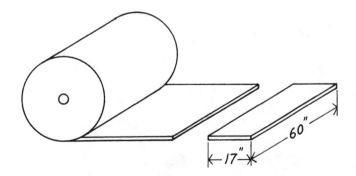

13. Illustrated is one of eight bundles of round bar stock which make up a delivery weighing 14,208 lbs. What is the weight of each bar?

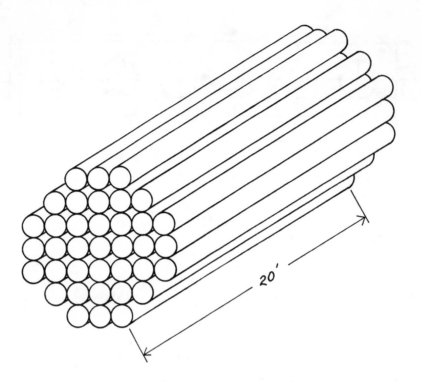

14. Studs are to be welded to this plate in the following manner. Studs along the length are to be spaced 9 in. apart and studs along the width are to be spaced 8 in. apart. How many studs are required?

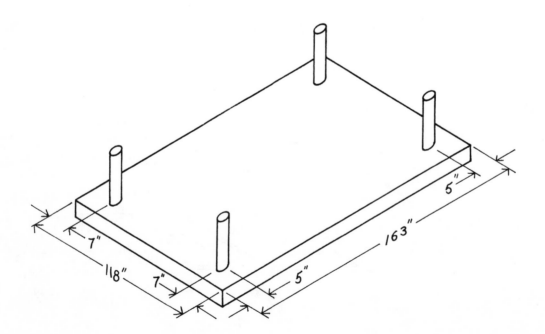

Section Two

COMMON FRACTIONS

Objectives for Section Two, Common Fractions

After studying this section, you will be able to:

■ Explain the parts of a fraction
■ Provide examples of proper fractions, improper fractions, and mixed numbers
■ Demonstrate how to reduce equivalent fractions
■ Demonstrate how to convert between mixed numbers and improper fractions
■ Perform addition of fractions
■ Perform subtraction of fractions
■ Perform multiplication of fractions
■ Perform division of fractions

Contents for Section Two, Common Fractions

INTRODUCTION TO COMMON FRACTIONS

INTRODUCTION

When you divide something, such as a steel rod, into a number of equal pieces, each of those pieces can be called a fraction of the whole rod. If the rod is cut into ten equal parts, each part is one-tenth of the whole. This can be expressed in numbers by the fraction 1/10. A fraction such as this is properly called a **common fraction.** In everyday usage, it is often referred to as a fraction. Three of the parts make up 3/10 of the whole, 9 parts would be 9/10, and finally, all ten parts could be written as 10/10. Since all ten parts make up the whole, you can see that 10/10 = 1. Any fractions containing all the parts are equal to 1: 64/64, 17/17, 256/256, 1/1, 12/12.

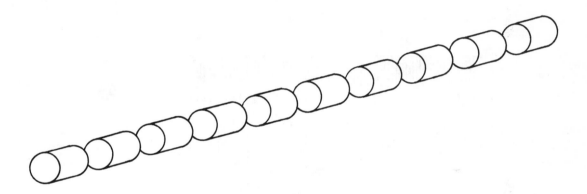

The following three illustrations show fractional parts of an inch.

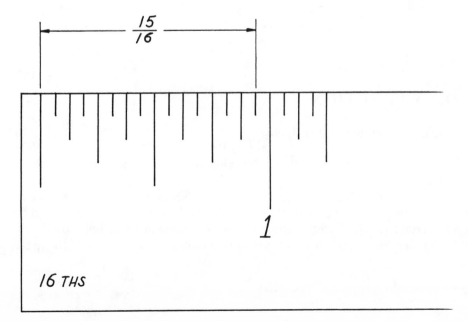

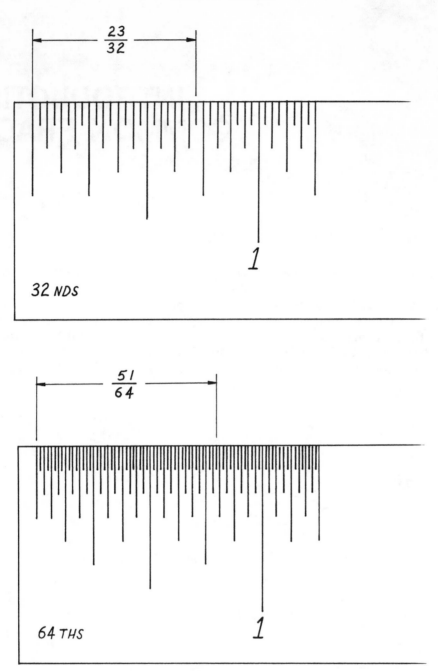

PARTS OF A FRACTION

The parts of a fraction are shown below.

$$\frac{3}{4} \begin{array}{l} \leftarrow \textbf{numerator} \\ \leftarrow \textbf{denominator} \end{array}$$

An easy way to remember these terms and to keep them separate is to think that the denominator is down on the bottom; a "d" in denominator and a "d" in down. The numerator and denominator together are referred to as the **terms** of the fraction. The **denominator** tells you how many equal parts the item has been divided into. The **numerator** tells you how many of those parts you are using.

PROPER FRACTIONS

Proper fractions have a numerator smaller than the denominator. Examples are: 1/4, 18/57, 9/32.

IMPROPER FRACTIONS

Improper fractions have a numerator larger than the denominator. They are "top heavy," such as: 15/7, 29/17, 9/8.

MIXED NUMBERS

Mixed numbers consist of whole numbers and fractions. Examples are: 1 1/2, 37 11/16, 108 5/13.

EQUIVALENT FRACTIONS

Equivalent fractions are fractions which are equal in value to each other. Any fraction can be expressed as an equivalent fraction in **higher terms** or **lower terms.**

EQUIVALENT FRACTIONS IN HIGHER TERMS

Multiply the numerator and denominator of a fraction by the same number and the fraction will then be expressed in higher terms. The value of the two fractions will be the same, as illustrated:

$$\frac{3 \text{ multiplied by } 7}{4 \text{ multiplied by } 7} = \frac{21}{28}$$

The fractions 3/4 and 21/28 are equivalent fractions.

EQUIVALENT FRACTIONS IN LOWER TERMS

Expressing a fraction in lower terms is called **reducing.** If you divide both the numerator and denominator of a fraction by the same number, the fraction will be reduced. The value of the fraction will not be changed. The numbers, of course, must divide evenly (that is, without a remainder) as illustrated:

$$\frac{24 \text{ divided by } 6}{30 \text{ divided by } 6} = \frac{4}{5}$$

The fractions 24/30 and 4/5 are equivalent fractions.

Here is another example:

$$\frac{\cancel{42}^{6}}{\cancel{56}_{8}} = \frac{\cancel{6}^{3}}{\cancel{8}_{4}} = \frac{3}{4}$$

When a fraction cannot be reduced further, it is considered to be in its **lowest terms.** When you arrive at the final answer, you are expected to reduce the answer to its lowest terms.

REDUCING IMPROPER FRACTIONS

Improper fractions are reduced by dividing the numerator by the denominator. If there is a remainder, it is expressed as a fraction, such as:

$$\frac{19}{4} = 4\frac{3}{4}$$

$$\frac{22}{10} = 2\frac{2}{10} = 2\frac{1}{5}$$

$$\frac{35}{5} = 7$$

CHANGING MIXED NUMBERS TO IMPROPER FRACTIONS

This is an operation useful in solving certain math problems you will encounter. Multiply the whole number by the denominator and then add the numerator.

$$7\frac{2}{3} = \frac{23}{3}$$

KEY TERMS FOR WELDERS
COMMON FRACTION
NUMERATOR
DENOMINATOR
TERMS
PROPER FRACTION
IMPROPER FRACTION
MIXED NUMBER
EQUIVALENT FRACTION
HIGHER TERMS
REDUCING
LOWER TERMS
LOWEST TERMS

Unit 6—Practicing with Common Fractions

Show all your work. Box your answers.

1. Express the following as equivalent fractions.

   (a) $\frac{5}{8} = \frac{?}{16}$ (b) $\frac{1}{2} = \frac{?}{200}$ (c) $\frac{1}{1} = \frac{?}{98}$

(d) $\dfrac{20}{64} = \dfrac{?}{16}$ (e) $\dfrac{28}{32} = \dfrac{?}{8}$ (f) $\dfrac{128}{256} = \dfrac{?}{2}$

2. Are the following pairs of fractions equivalent fractions?

   (a) $\dfrac{2}{3}$ and $\dfrac{65}{99}$ (b) $\dfrac{3}{4}$ and $\dfrac{21}{28}$

   (c) $\dfrac{17}{64}$ and $\dfrac{17}{128}$ (d) $\dfrac{1}{1}$ and $\dfrac{2}{1}$

   (e) $\dfrac{10}{11}$ and $\dfrac{110}{121}$ (f) $\dfrac{19}{33}$ and $\dfrac{76}{132}$

3. Reduce the following improper fractions.

   (a) $\dfrac{18}{4}$ (b) $\dfrac{111}{11}$ (c) $\dfrac{141}{8}$

   (d) $\dfrac{1968}{16}$ (e) $\dfrac{7}{2}$ (f) $\dfrac{75}{25}$

4. Change the following mixed numbers to improper fractions.

   (a) $1\dfrac{1}{2}$ (b) $12\dfrac{3}{4}$ (c) $17\dfrac{11}{64}$

   (d) $29\dfrac{1}{8}$ (e) $37\dfrac{3}{7}$ (f) $185\dfrac{5}{9}$

5. Express the following fractions in the lowest terms.

   (a) $\dfrac{2}{4}$ (b) $\dfrac{30}{36}$ (c) $\dfrac{98}{112}$

   (d) $\dfrac{12}{108}$ (e) $\dfrac{64}{656}$ (f) $\dfrac{16}{2048}$

ADDITION OF FRACTIONS

INTRODUCTION

The basic operations of addition, subtraction, multiplication, and division are also performed with fractions and combinations of fractions and whole numbers. Handling these calculations is probably one of the most fundamental math jobs a welder is called upon to do.

METHOD USED TO ADD FRACTIONS

Three main situations will be encountered when adding fractions; fractions with common denominators, fractions without common denominators, and mixed numbers without common denominators.

ADDING FRACTIONS WITH COMMON DENOMINATORS

Fractions can be added only if they have the same denominators. For example, 3/11 and 4/11 have a **common denominator.** To add them, just add the numerators (3 + 4 = 7) and place this answer over the common denominator (7/11).

ADDING FRACTIONS WITHOUT COMMON DENOMINATORS

Fractions without common denominators can be added by first changing the denominators so they are the same. The whole operation is easier if the denominator selected is the lowest one possible. Your first objective, then, is to find the **lowest common denominator (L.C.D.).** An efficient way of doing this is described below.

1. Add the following fractions:

$$\frac{5}{6} + \frac{3}{8} + \frac{2}{3}$$

2. Write all denominators in a row. It is a good idea to separate them by commas so they do not blend together.

$$6, \quad 8, \quad 3$$

3. Find a number which divides evenly into at least two of the numbers listed. In this example, 2 will work. Divide using the format illustrated.

$$2 \overline{)6, \quad 8, \quad 3}$$
$$3, \quad 4, \quad 3$$

Here is what happened. Two divided evenly into 6 three times, so 3 was written down. Two divided evenly into 8 four times, so 4 was written down. Then, 2 did not divide evenly into 3, so the 3 was "brought down."

4. The next step follows the pattern of the previous step. This time, inspect the new line of 3, 4, 3 to find a number that divides evenly into at least two of the numbers. Obviously, the number is 3.

$$
\begin{array}{r}
2\,)\underline{6,\quad 8,\quad 3} \\
3\,)\underline{3,\quad 4,\quad 3} \\
1,\quad 4,\quad 1
\end{array}
$$

At this point, the line of digits cannot be reduced further.

5. Now multiply the numbers 2, 3, 1, 4, 1:
$$2 \times 3 \times 1 \times 4 \times 1 = 24$$

The L.C.D. is 24.

6. The final step is to express each fraction in the original problem as an equivalent fraction with denominator 24, and then add.

$$\frac{5}{6} + \frac{3}{8} + \frac{2}{3} = \frac{20}{24} + \frac{9}{24} + \frac{16}{24} = \frac{45}{24}$$

$$\frac{45}{24} = 1\frac{21}{24} = 1\frac{7}{8}$$

ADDING MIXED NUMBERS WITHOUT COMMON DENOMINATORS

There are several ways to add mixed numbers without common denominators. The most commonly used method is explained next.

1. Add $2\frac{3}{4} + 5\frac{1}{7}$

2. Rewrite the question in this manner.

$$
\begin{array}{r}
2\frac{3}{4} \\
+\ 5\frac{1}{7} \\
\hline
\end{array}
$$

3. Find the L.C.D. and write as equivalent fractions.

$$
\begin{array}{r}
2\frac{12}{28} \\
+\ 5\frac{4}{28} \\
\hline
\end{array}
$$

4. Add the whole numbers and then the fractional parts.

$$
\begin{array}{r}
2\frac{12}{28} \\
+\ 5\frac{4}{28} \\
\hline
7\frac{16}{28}
\end{array}
$$

5. Reduce the answer to its lowest terms.

$$7\frac{16}{28} = 7\frac{4}{7}$$

KEY TERMS FOR WELDERS
LOWEST COMMON DENOMINATOR L.C.D. COMMON DENOMINATOR

Unit 7—Practicing Addition of Fractions

Show all your work. Box your answers.

1. Add the following:

   (a) $\dfrac{3}{18} + \dfrac{7}{18} + \dfrac{1}{18} =$

   (b) $\dfrac{21}{96} + \dfrac{13}{96} + \dfrac{37}{96} =$

   (c) $\dfrac{109}{219} + \dfrac{25}{219} + \dfrac{34}{219} =$

   (d) $\dfrac{3}{5} + \dfrac{4}{15} =$

   (e) $\dfrac{17}{25} + \dfrac{4}{5} + \dfrac{49}{50} =$

   (f) $\dfrac{15}{28} + \dfrac{3}{4} + \dfrac{6}{7} =$

2. Add the following:

   (a) $\dfrac{3}{4} + \dfrac{1}{3} + \dfrac{4}{5} + \dfrac{1}{2} =$

   (b) $\dfrac{3}{8} + \dfrac{3}{7} + \dfrac{7}{12} =$

   (c) $\dfrac{107}{120} + \dfrac{85}{150} =$

(d) $\dfrac{7}{27} + \dfrac{31}{36} + \dfrac{3}{23} =$

(e) $\dfrac{7}{64} + \dfrac{41}{88} + \dfrac{53}{92} =$

(f) $\dfrac{117}{365} + \dfrac{99}{200} + \dfrac{87}{260} =$

3. Add the following:

(a) $13\dfrac{1}{4} + 5\dfrac{5}{8} + 4\dfrac{9}{16} =$

(b) $\dfrac{3}{56} + \dfrac{3}{16} + \dfrac{7}{28} + \dfrac{2}{7} =$

(c) $38\dfrac{19}{34} + 119\dfrac{11}{12} =$

(d) $11\dfrac{1}{4} + 27\dfrac{3}{16} + \dfrac{7}{8} + 10\dfrac{1}{2} =$

(e) $\dfrac{25}{140} + \dfrac{9}{204} =$

(f) $1\dfrac{109}{187} + \dfrac{75}{121} =$

4. According to their time cards, the following people spent the indicated amount of time on Job #8815.

JOB #8815	
NAME	HOURS
Lee Jones	44 1/4
Jamil El Helou	185 3/4
Filomena Ferrari	121 1/2
Pat Anderson	15
Luis Martin	29 1/4
Guy Laframbois	6 1/2

Addition of Fractions

What was the total amount of time spent on the job?

5. A stainless steel rod, one-half inch in diameter, has been cut into five pieces of the following lengths:

 —twenty-one and three-quarter inches;
 —seven inches;
 —thirty-two and seven-thirty-second inches;
 —eleven and one-half inches;
 —sixteen and fifteen-sixteenth inches.

What was the original length of the rod? Ignore losses due to cutting.

6. A welder has to cut a 1 in. square rod into three pieces of the following lengths: 23 5/8 inches, 26 1/4 inches and 17 5/16 inches. Each cut will waste 1/4 inch of material. The pieces are cut from a 96 inch rod. What length of the rod is used to make the three pieces?

7. What is the overall length of a machine part consisting of four pieces measuring 5/8 inches, 43 9/16 inches, 12 27/32 inches, 15/64 inches?

8. What is the outside diameter of a machined brass tube with an inside diameter of 22 7/8 in. and a thickness of 13/32 in.?

9. The individual parts of a weldment had the following weights: 135 1/2 lbs., 19 3/16 lbs., 71 1/4 lbs., 6 3/8 lbs., 1 1/4 lbs. The filler material used in welding added an additional 2 3/32 lbs. What is the final weight of the completed weldment?

10. Find the distance from stud (A) to stud (B).

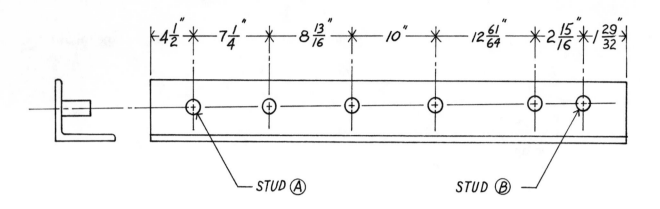

11. Calculate the overall length.

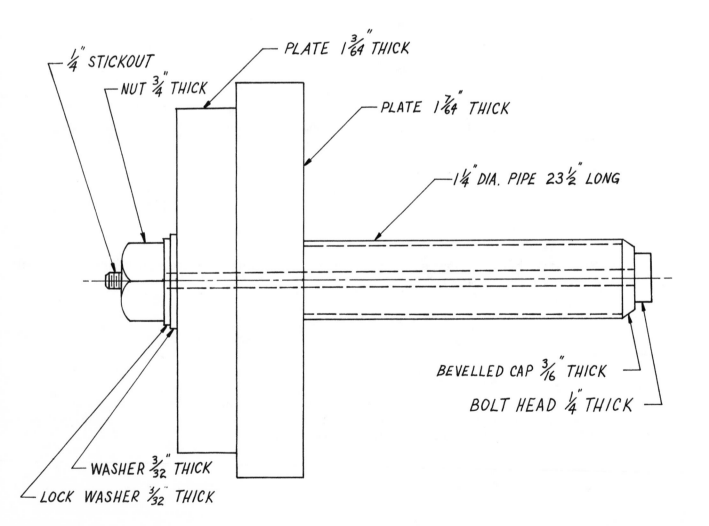

12. Calculate the following lengths.
    (a) The total of the vertical lengths.
    (b) The total of the horizontal lengths.
    (c) The total of the angular lengths.

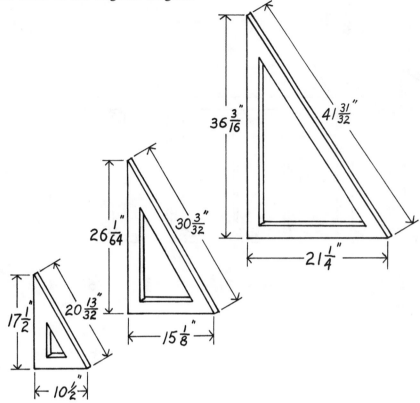

13. Calculate the overall length of this test bar after welding.

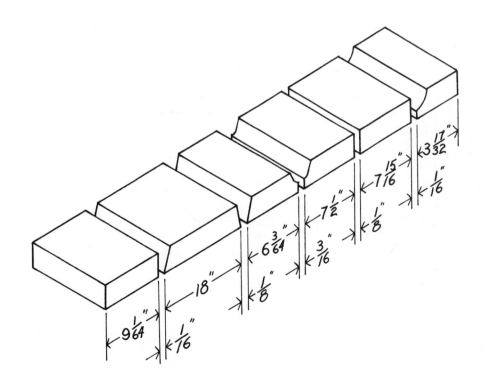

14. (a) What is the distance from hanger (A) to hanger (B)?
    (b) What is the total length of the weldment?

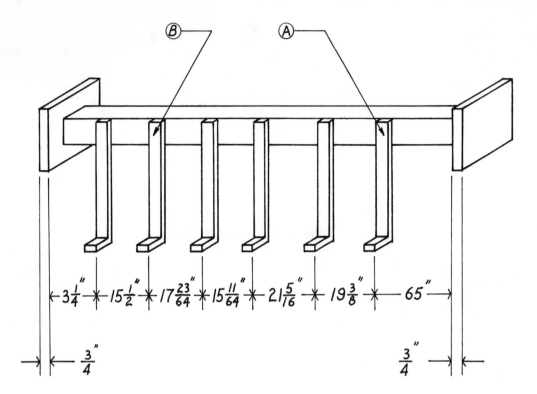

15. What is the total length of angle iron used in welding the following frame?

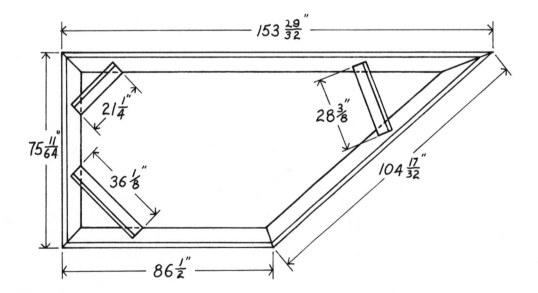

SUBTRACTION OF FRACTIONS

INTRODUCTION

Subtracting fractions is very similar to adding fractions. You will see the method is generally the same except, naturally, that a subtraction is performed rather than an addition.

METHOD USED TO SUBTRACT FRACTIONS

Three main situations will be encountered where you will have to subtract fractions: fractions with common denominators, fractions without common denominators, and mixed numbers without common denominators.

SUBTRACTING FRACTIONS WITH COMMON DENOMINATORS

Fractions with common denominators are easily subtracted. You simply subtract the numerators and place the result over the common denominator.

$$\frac{9}{17} - \frac{3}{17} = \frac{6}{17}$$

SUBTRACTING FRACTIONS WITHOUT COMMON DENOMINATORS

Fractions without common denominators can be subtracted by first changing the denominators so they are the same. As explained in Unit 7, finding the L.C.D. is usually the easiest way to do this. Review that explanation, if necessary. Then study the following example:

$$\frac{13}{14} - \frac{6}{63} =$$

$$7\,\overline{)14,\ \ 63}$$
$$\quad\ \ 2,\ \ \ 9$$

$$7 \times 2 \times 9 = 126$$

$$\text{L.C.D.} = 126$$

Therefore:

$$\frac{13}{14} - \frac{6}{63} = \frac{117}{126} - \frac{12}{126} = \frac{105}{126}$$

SUBTRACTING MIXED NUMBERS WITHOUT COMMON DENOMINATORS

To subtract mixed numbers without common denominators, first change the fractional parts to common denominators. Then subtract the fractional parts. Finally, subtract the whole numbers.

$$17\frac{3}{4} = 17\frac{12}{28}$$
$$-\ 4\frac{1}{7} = -\ 4\frac{4}{28}$$
$$13\frac{8}{28} = 13\frac{2}{7}$$

Not all subtractions will follow these exact steps. A complication will occur when the fractional part to be subtracted is larger than the fraction from which it is being subtracted. When this occurs, borrow 1 from the whole number and add it to the fraction part of the number. The borrowed 1 is first changed to an appropriate fraction, such as 3/3, 8/8, 16/16, etc. (see Unit 6).

$$11\frac{9}{16} = 11\frac{9}{16} = 10\frac{25}{16}$$
$$-\ 5\frac{3}{4} = -\ 5\frac{12}{16} = -\ 5\frac{12}{16}$$

One was borrowed from 11 and added to 9/16 as:

$$\frac{9}{16} + \frac{16}{16} = \frac{25}{16}$$

Now, to finish the problem:

$$10\frac{25}{16}$$
$$-\ 5\frac{12}{16}$$
$$5\frac{13}{16}$$

Unit 8—Practicing Subtraction of Fractions

1. Subtract the following:

(a) $\frac{5}{9} - \frac{2}{9} =$

(b) $\frac{13}{16} - \frac{5}{16} =$

(c) $\frac{62}{101} - \frac{31}{101} =$

(d) $\frac{7}{8} - \frac{1}{4} =$

(e) $\dfrac{15}{16} - \dfrac{7}{8} =$

(f) $35\dfrac{17}{20} - 25\dfrac{11}{17} =$

2. Subtract the following:

(a) Reduce 738 7/32 by 48 9/32.

(b) Find the difference between 6499 3/4 and 6460 1/2.

(c) 111/119 less 111/238.

(d) Subtract 19/64 from 5/8.

(e) Take 49 2/3 away from 65.

(f) Calculate the difference between 2020 1/2 and 1100 1/2.

3. Subtract the following:

(a) 7 1/3 − 6 2/3 =

(b) 11 9/16 − 7/8 =

(c) 1 3/10 − 5/6 =

(d) 19 2/5 − 18 9/10 =

(e) 45 1/3 − 32 2/3 =

(f) 24 1/4 − 15 2/3 =

4. A piece of steel (U.S. Standard gage #28), 3 ft.-8 1/4 in. long was sheared from a sheet 9 ft.-11 3/16 in. long. What length of the original sheet remains?

5. The sides of a square piece of steel measure 18 5/8 in. Using two cuts, the piece is sheared to 11 3/64 in. × 14 3/16 in. What is the width of each removed piece?

6. Eighty-seven and one-half miles of pipeline were to be inspected by Twin City Consulting Inc. The company estimated they could inspect 5 1/8 miles of line per week. What length of pipeline would remain to be inspected after: one week? two weeks? three weeks?

7. In Detroit's Industrial Softball League, halfway through the season the Pipefitters were 17 games behind the leader and in 5th place. The Welders were in 4th place and 13 1/2 games behind the leader. How many games behind the Welders were the Pipefitters?

8. The wall thickness of a piece of round tubing is 3/16 in. and the outside diameter is 3 1/4 in. What is the inside diameter?

9. Four pieces measuring 23 7/32 in., 18 in., 19 1/2 in., and 22 1/4 in. are to be cut from a stock piece of 1 in. x 1 in. square tubing that is 127 1/4 in. long. Each saw cut wastes 3/32 in. of material. What will be the length of the stock piece after cutting?

10. What is the length of piece removed when each of the following are reduced to the length indicated? The saw cut in each case will waste 1/16 in. of material.
    (a) This 1/4 in. square bar is reduced to 19 1/2 in.

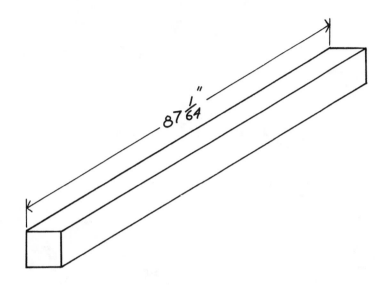

(b) This 1 1/4 in. × 1 1/4 in. square tubing is reduced to 32 5/8 in.

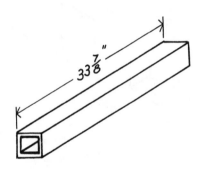

(c) This 2 in. channel is reduced to 41 7/8 in.

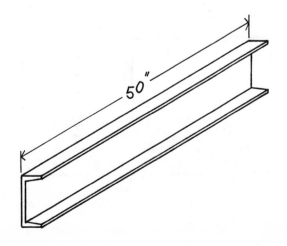

(d) This 3/4 in. round bar is reduced to 39 in.

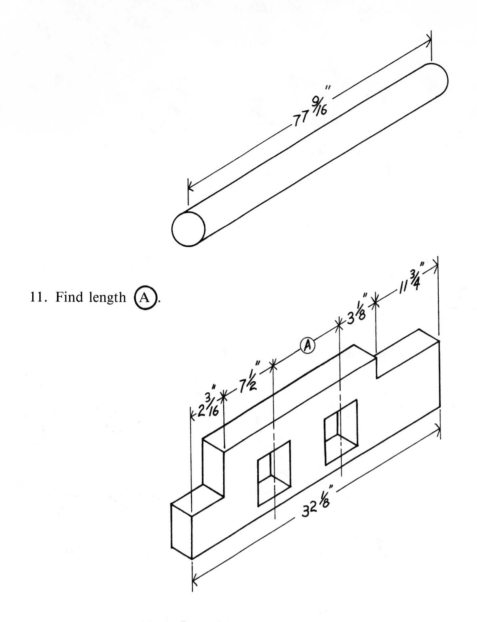

11. Find length Ⓐ.

12. Find lengths Ⓐ, Ⓑ, Ⓒ.

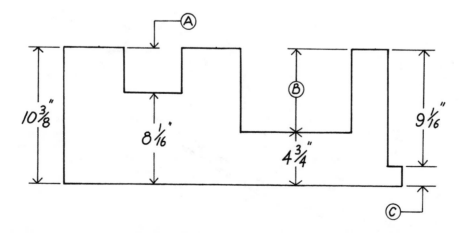

13. Slots are cut from the circular piece as shown. Calculate dimension (A) and dimension (B).

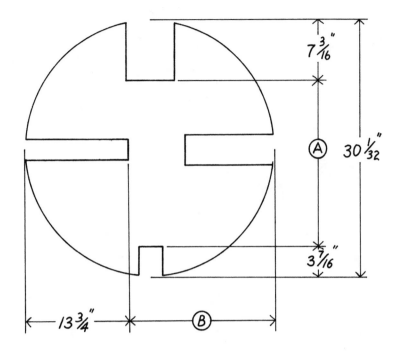

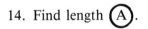

14. Find length (A).

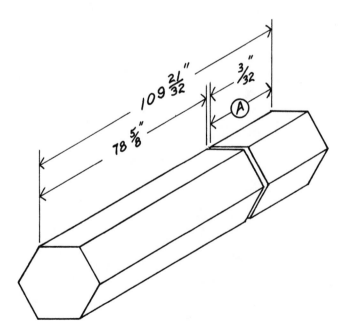

15. Three pieces measuring 18 11/64 in., 9 3/8 in., 82 1/16 in. are cut from this half-round stock. Each saw cut wastes 3/32 in. of material. What is the length of the remaining piece?

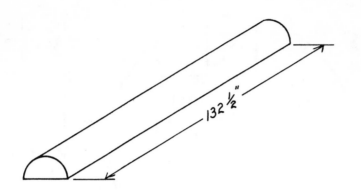

$132\frac{1}{2}''$

Unit 9

MULTIPLICATION OF FRACTIONS

INTRODUCTION

It may seem surprising, but multiplying fractions is easier than adding or subtracting fractions. This is because the denominators do not have to be the same. You do not have to find a common denominator.

Three expressions are commonly used to indicate multiplication of fractions.

$$1/5 \times 3/4$$
$$1/5 \text{ times } 3/4$$
$$1/5 \text{ of } 3/4$$

The one expression that may seem peculiar is, "of," however, through common usage, this word has come to mean multiplication.

METHOD USED TO MULTIPLY FRACTIONS

To multiply $3/4 \times 5/7$, multiply the numerators and then multiply the denominators.

$$\frac{3}{4} \times \frac{5}{7} = \frac{15}{28}$$

As always, if the answer can be reduced, you would do so. Sometimes a question can be reduced to a simpler form before beginning the multiplication. You can do this by finding a number that will divide evenly into *any* of the numerators and denominators. Continue doing this until the numerators and denominators cannot be reduced any further.

$$\frac{\cancel{16}^{4}}{\cancel{100}_{4}^{1}} \times \frac{\cancel{25}^{1}}{37} = \frac{4}{37}$$

You will find that multiplying fractions is much easier if you first search for cases where you can reduce the fractions before such divisions.

MULTIPLYING MIXED NUMBERS

Probably the easiest way to multiply mixed numbers is to first change them to improper fractions, as shown here:

$$3\frac{1}{2} \times 9\frac{2}{5}$$

$$\frac{7}{2} \times \frac{47}{5} = \frac{329}{10} = 32\frac{9}{10}$$

If the task is to multiply a mixed number and a whole number, rewrite the whole number as a fraction with a denominator of 1; for example, 42 is rewritten as 42/1. A typical problem would look like this:

$$42 \times 2\frac{4}{7}$$

$$\frac{\overset{6}{\cancel{42}}}{1} \times \frac{18}{\cancel{7}_1} = \frac{108}{1} = 108$$

MULTIPLYING MORE THAN TWO FRACTIONS

Multiplying more than two fractions follows the same routine as multiplying two fractions. First, try to reduce the question to a simpler form and then multiply the numerators and denominators.

$$\frac{1}{2} \times \frac{\overset{1}{\cancel{3}}}{4} \times \frac{5}{\underset{2}{\cancel{6}}} = \frac{5}{16}$$

$$5\frac{1}{2} \times 4\frac{2}{3} \times 9\frac{3}{5}$$

$$\frac{11}{\cancel{2}_1} \times \frac{\overset{7}{\cancel{14}}}{\cancel{3}_1} \times \frac{\overset{16}{\cancel{48}}}{5} = \frac{1232}{5} = 246\frac{2}{5}$$

Unit 9—Practicing Multiplication of Fractions

Show all your work. Box your answers.

1. Multiply the following:

(a) $\frac{1}{5} \times \frac{1}{3} =$ (b) $\frac{3}{4} \times \frac{7}{15} =$

(c) $\frac{15}{33} \times \frac{12}{61} =$ (d) $\frac{1}{2} \times \frac{1}{2} =$

(e) $\frac{13}{64} \times \frac{8}{9} =$ (f) $\frac{7}{35} \times \frac{12}{144} =$

2. Multiply the following:

   (a) $4 \times \dfrac{3}{16} =$

   (b) $3\dfrac{1}{3} \times 5\dfrac{1}{2} =$

   (c) $17\dfrac{2}{5} \times 2\dfrac{1}{4} =$

   (d) $8\dfrac{1}{2} \times \dfrac{15}{45} =$

   (e) $9\dfrac{3}{8} \times 9\dfrac{3}{8} =$

   (f) $15 \times \dfrac{15}{16} =$

3. Multiply the following:

   (a) $16/27 \times 3/4 \times 5/7 =$

   (b) $3\ 1/2 \times 4\ 1/6 \times 1\ 3/10 =$

   (c) $11\ 3/16 \times 1\ 1/8 \times 2 =$

   (d) $23/140 \times 70/92 \times 17/33 =$

   (e) $7 \times 1/2 \times 3/4 =$

   (f) $9 \times 4/15 \times 9\ 4/15 =$

4. A pipeline is laid at the rate of 1/8 mile per day. How many miles of line would be completed in 29 1/2 days?

5. One cubic foot of water weighs 62 1/2 lbs. Find the weight of the contents of a welded steel tank containing 28 cubic feet of water.

6. Last year, a fabricating shop lost 135 1/2 man hours of labor due to accidents. This year, they lost 1 1/2 times that amount. How many hours were lost this year?

7. Two grain hoppers were built by a crew consisting of three welders who worked on the job 4 1/4 hours a day for 17 days. How many hours did the crew work on the hoppers?

8. A GMAW welder running at a speed of 16 3/8 in. per minute requires 14 1/2 minutes to complete a V-groove weld on an aluminum sill. What is the length of the sill?

9. During each hour, a shop uses the following quantities of weld material:

     135 1/8 lbs. of 3/16 in. wire
     83 3/16 lbs. of 5/64 in. wire
     27 3/4 lbs. of 1/4 in. wire

   (a) How many pounds of 1/4 in. wire are used in 5 1/15 hours?

   (b) How many pounds of 3/16 in. wire are used in 8 1/2 hours?

   (c) How many pounds of 5/64 in. wire are used in 3 7/60 hours?

10. The following weldment requires 6 2/3 rods. If 40 1/2 weldments are completed in 2 1/4 days, how many rods will be required?

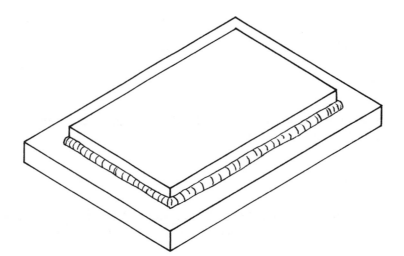

11. Nine pillars are to be fabricated. For the purpose of weight reduction, each pillar is to have 13 holes flame cut in the center plate. Seven large holes are to be cut, reducing the weight by 32 19/32 lbs. per hole. Six smaller holes are to be cut, reducing the weight by 23 7/16 lbs. per hole. What is the total weight reduction for the entire project?

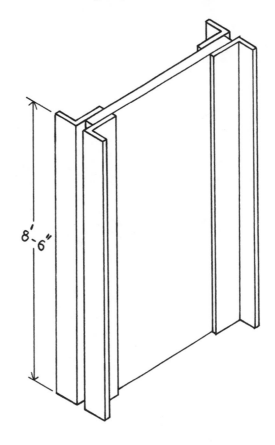

12. A flame cutting job using Pattern #1 resulted in 3/16 lbs. of scrap per part. The scrap rate for Pattern #2 was 2/5 of the rate for Pattern #1.
    (a) What weight of scrap would result if Pattern #1 is used to produce 5,040 parts?
    (b) What weight of scrap would result if Pattern #2 is used to produce the same number of parts?

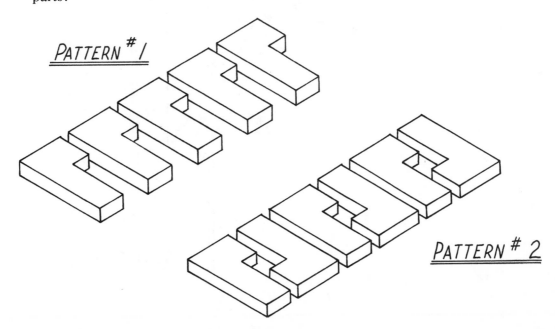

13. One part, as illustrated below, is produced every 47/60 of a minute.
    (a) How many minutes are required to produce 2,864 parts?
    (b) What is the total length of weld deposited for 2,864 parts?

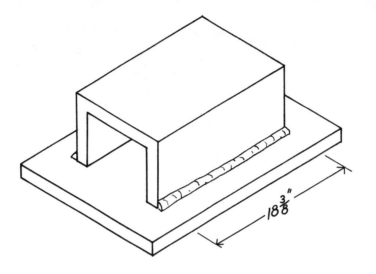

14. The structural work for a new shopping mall included 1,047 hangers.
    (a) What is the total length of 3" x 2" angle used?
    (b) What is the total length of 1" x 1" angle used?

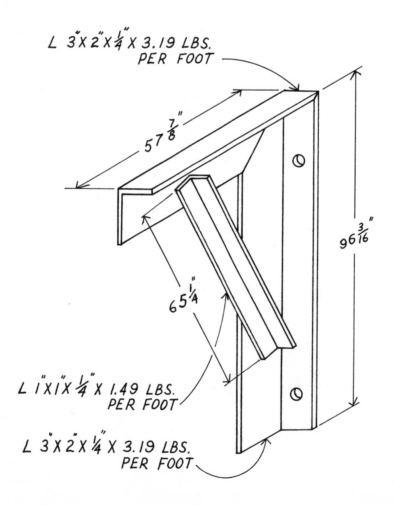

15. Calculate the weight of the weldment.

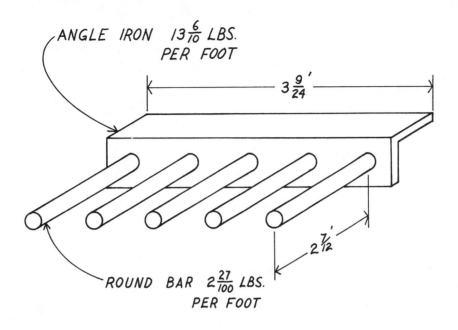

ANGLE IRON $13\frac{6}{10}$ LBS. PER FOOT

$3\frac{9}{24}'$

ROUND BAR $2\frac{27}{100}$ LBS. PER FOOT

$2\frac{7}{12}'$

DIVISION OF FRACTIONS

INTRODUCTION

Learning to divide fractions is quite easy once you have learned to multiply fractions. You will see why as you read this unit.

METHOD USED TO DIVIDE FRACTIONS

To divide, **invert the divisor** (that is, turn the divisor upside down) and change the operation from a division to a multiplication. Remember, the divisor is the number doing the division. In the following example, 3/5 is being divided by 5/8, the divisor.

$$\frac{3}{5} \div \frac{5}{8} = \frac{3}{5} \times \frac{8}{5} = \frac{24}{25}$$

It may seem odd that a division problem can suddenly be switched to a multiplication problem. Although they are opposite operations, multiplication and division are closely related. Because of this close relationship it is possible to convert fractional division problems to multiplications as shown above.

DIVIDING MIXED NUMBERS

As in multiplying fractions, first change the mixed numbers to improper fractions. Then, invert the divisor and multiply. As usual, try to reduce the problem to a simpler form before proceeding with the multiplication part of the problem. Here is an example:

$$7\frac{3}{4} \div 7\frac{1}{2} = \frac{31}{4} \div \frac{15}{2} = \frac{31}{\overset{}{\underset{2}{4}}} \times \frac{\overset{1}{\cancel{2}}}{15} = \frac{31}{30} = 1\frac{1}{30}$$

COMPLEX FRACTIONS

There are several ways to symbolize division of fractions. Here are two different formats you can expect to encounter:

The standard format:

$$\frac{13}{16} \div \frac{3}{4}$$

The complex fraction format:

$$\frac{\dfrac{5}{8}}{\dfrac{2}{3}}$$

A **complex fraction** has a fraction for the numerator and a fraction for the denominator. To perform this division simply, rewrite it in the standard format and proceed in the usual manner:

$$\frac{\frac{5}{8}}{\frac{2}{3}} = \frac{5}{8} \div \frac{2}{3} = \frac{5}{8} \times \frac{3}{2} = \frac{15}{16}$$

So, you can see that even though a complex fraction may appear at first very difficult to divide, it really is quite simple. The same general procedure is followed when the numerator and/or denominator contain mixed numbers:

$$\frac{\frac{9}{144}}{6\frac{3}{4}} = \frac{9}{144} \div 6\frac{3}{4} = \frac{9}{144} \div \frac{27}{4}$$

$$= \frac{9}{144} \div \frac{27}{4} = \frac{\overset{1}{\cancel{9}}}{\underset{36}{\cancel{144}}} \times \frac{\overset{1}{\cancel{4}}}{\underset{3}{\cancel{27}}} = \frac{1}{108}$$

KEY TERMS FOR WELDERS
INVERT THE DIVISOR COMPLEX FRACTION

Unit 10—Practicing Division of Fractions

Show all your work. Box your answers.

1. Divide the following:

   (a) $\dfrac{1}{3} \div \dfrac{5}{16} =$

   (b) $\dfrac{4}{5} \div \dfrac{9}{10} =$

   (c) $\dfrac{5}{6} \div \dfrac{7}{8} =$

   (d) $\dfrac{7}{11} \div \dfrac{7}{11} =$

   (e) $\dfrac{3}{4} \div \dfrac{4}{3} =$

   (f) $\dfrac{1}{2} \div \dfrac{1}{4} =$

2. Divide the following:

(a) $21 \div \dfrac{5}{6} =$

(b) $2\dfrac{5}{16} \div 13 =$

(c) $57 \div 2\dfrac{5}{8} =$

(d) $63 \div \dfrac{1}{63} =$

(e) $9\dfrac{3}{16} \div 3\dfrac{9}{16} =$

(f) $100 \div \dfrac{1}{2} =$

3. Divide the following:

(a) $1\dfrac{13}{16} \div 2\dfrac{1}{4} =$

(b) $5\dfrac{4}{9} \div 5\dfrac{4}{9} =$

(c) $\dfrac{7\dfrac{9}{16}}{7} =$

(d) $\dfrac{4\dfrac{2}{5}}{2\dfrac{3}{4}} =$

(e) $\dfrac{3\dfrac{5}{6}}{9\dfrac{1}{5}} =$

(f) $\dfrac{\dfrac{123}{716}}{\dfrac{82}{182}} =$

4. How many 3 1/2 in. pieces can be sheared from a thin piece of sheet metal 31 in. long?

5. A piece of 1/2 in. diameter copper tubing 105 1/2 in. long is cut with a pipe cutter into 6 pieces of equal length. What is the length of each piece?

6. A bus, rented by the shop, is scheduled to deliver a crew of workers to a job site 87 1/2 miles away. The bus departs from the shop at 6:30 a.m. and is expected to arrive at 9:00 a.m. (a) What average speed must be maintained to arrive on schedule?

   (b) If the bus averages 21 1/2 miles per gallon, how many gallons of gas are needed for one trip to the site and back to the shop?

7. A sheet of metal weighs 18 7/16 lbs. In a shearing operation, the sheet is cut into strips weighing 3/8 lbs. each. How many strips of metal are produced?

8. A steel plate 83 1/2 in. long weighs 233 4/5 lbs. How much does a 1 in. length of the plate weigh?

9. A volume of 1 cubic foot contains about 7 1/2 gallons. How many cubic feet of oil will a cooling tank hold if it contains 2,770 1/2 gallons?

10. Divide this T-shape into 13 equal parts. What length are the pieces?

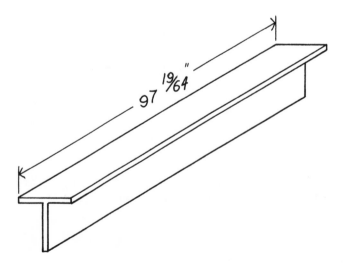

11. The following square bar of hot rolled steel is cut into four equal length pieces. Each saw cut wastes 3/32 in. of material. What length are the pieces?

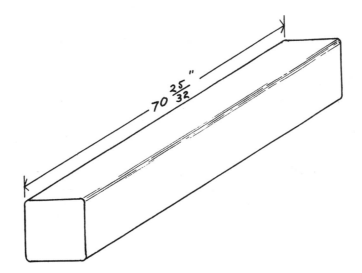

12. The cross section of a piece of extra strong pipe has the following dimensions. What is the wall thickness of the pipe?

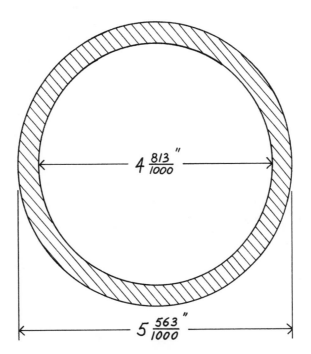

13. How many strips 48 in. long and 13 7/16 in. wide can be sheared from this sheet of 15 gage steel?

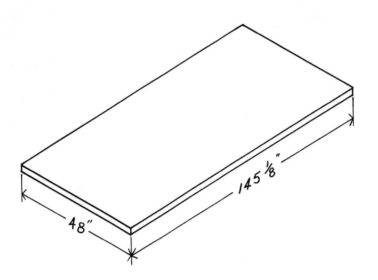

14. How many 7 3/4 in. square pieces can be cut from this sheet?

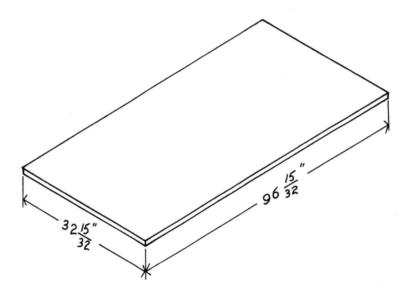

15. Fifteen more pieces of tubing are to be welded equal distances apart. Determine the distance between the pieces.

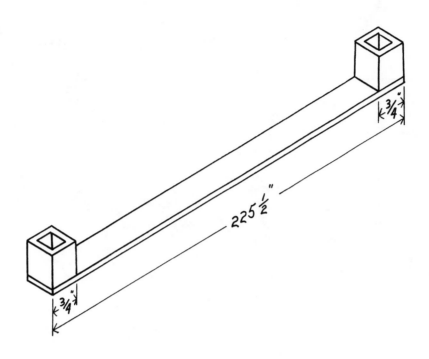

Section Three

DECIMAL FRACTIONS

Objectives for Section Three, Decimal Fractions

After studying this section, you will be able to:

- Give examples of decimal fractions
- Explain how to convert decimal fractions and common fractions
- Show how to round decimal fractions
- Perform addition of decimal fractions
- Perform subtraction of decimal fractions
- Perform multiplication of decimal fractions
- Perform division of decimal fractions

Contents for Section Three, Decimal Fractions

Unit 11

INTRODUCTION TO DECIMAL FRACTIONS

INTRODUCTION

As you have seen so far, numbers can be classified in a variety of ways, such as whole numbers, common fractions, mixed numbers, etc. This unit introduces another classification of numbers called decimal numbers.

In Unit One, it was noted that our numbering system is based on ten digits and that larger numbers are created by lining up these digits in a certain order. The decimal system uses this method of place position values to express numbers that are <u>less</u> than whole numbers. These are called **decimal fractions.**

All decimal numbers include a **decimal point.** Digits to the left of the decimal point are whole numbers. Digits to the right of the decimal point are fractional numbers. Each position to the right of the decimal has a value. Listed below are some of the names and place values.

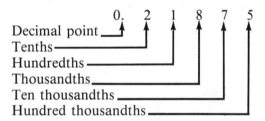

There are two ways of expressing decimal fractions verbally:
1. Pronounce each digit individually;
   14.63;
   "Fourteen point (or decimal) six three."
2. Pronounce the decimal fraction as a whole number
   and add the name of the last place value;
   14.63;
   "Fourteen and sixty three hundredths."
Always include a zero to the left of the decimal point if there are no whole numbers in the units place.

CONVERTING DECIMAL FRACTIONS TO COMMON FRACTIONS

A decimal fraction is changed to a common fraction by using the last place value as the denominator and the decimal fraction digits as the numerator:

$$0.3 = \frac{3}{10}$$

$$0.27 = \frac{27}{100}$$

$$0.173 = \frac{173}{1000}$$

81

The task is simplified by the fact that denominators will always be multiples of 10, such as 10, 100, 1,000, 10,000 etc. The actual value depends upon the place position of the last digit. Once you have done the conversion, you should reduce the fraction, if possible.

$$0.0625 \quad = \quad \frac{625}{10000} \quad = \quad \frac{25}{400} \quad = \quad \frac{1}{16}$$

$$0.828125 \quad = \quad \frac{828125}{1000000} \quad = \quad \frac{33125}{40000} \quad = \quad \frac{53}{64}$$

CONVERTING COMMON FRACTIONS TO DECIMAL FRACTIONS

A common fraction is changed to a decimal fraction by dividing the numerator by the denominator:

$$\frac{3}{4} \quad = \quad 4\overline{\smash{\big)}3}$$

Place a decimal to the right of the dividend. Place another decimal directly above that decimal. This is done so the decimal in the answer (the quotient) is properly located:

$$\frac{3}{4} \quad = \quad 4\overline{\smash{\big)}3.}$$

Now, add a zero to the right of the decimal in the dividend and begin to divide. Add zeros to the dividend as needed:

$$
\begin{array}{r}
.75 \\
4\overline{\smash{\big)}3.00} \\
\underline{2\,8} \\
20 \\
\underline{20} \\
0\ R
\end{array}
$$

ROUNDING DECIMALS

The previous example is typical of most fractions. However, there are two situations you should watch for and take special action. First, there are some common fractions which result in an unending decimal fraction. For example, 1/3 will convert to the unending decimal fraction 0.33333. Second, some common fractions will convert to decimal fractions that are quite long and much more accurate than the original common fraction. Therefore, it is common practice to reduce such numbers to a degree of accuracy that is adequate. This process is called **rounding the decimal** and is done as follows:

Determine the degree of accuracy required. Normally, the problem description will indicate the degree of accuracy required or to what place the decimal fraction should be extended. Also, the accuracy required is usually indicated as a tolerance on the shop drawings you will be using. An explanation of tolerance will be covered later in this text.

Eliminate all digits beyond the required degree of accuracy. If the first number you eliminate is 5 or more, increase the final number in your answer by 1. Following is an example:

1. Round 75.13846 to the nearest hundredth.

2. 75.13846.

3. Since 8, the first number eliminated, is greater than 5, increase the 3 to a 4.

4. The answer is 75.14.

KEY TERMS FOR WELDERS
DECIMAL FRACTION DECIMAL POINT ROUNDING THE DECIMAL

Unit 11—Practicing with Decimal Fractions

1. Write decimal numbers for the following:
   (a) One hundred decimal zero one.

   (b) Ninety five hundredths.

   (c) Fourteen decimal zero zero one two five.

   (d) Three thousand two hundred nineteen decimal one two five.

   (e) Decimal seven zero seven.

   (f) One thousand nine hundred seventeen ten thousandths.

   (g) Decimal eight six six.

   (h) Five decimal five one five six three.

2. Convert the following decimal fractions to common fractions:
   (a) 0.35 (b) 0.28125

   (c) 0.1665 (d) 0.6875

   (e) 0.3333 (f) 0.78125

3. Convert the following to decimal fractions:
   (a) $\dfrac{1}{2}$ (b) $10\dfrac{1}{10}$

   (c) $\dfrac{1}{32}$ (d) $\dfrac{45}{64}$

   (e) $\dfrac{7}{8}$ (f) $\dfrac{3}{1000}$

4. Round the following decimals as indicated:

To the nearest tenth:

(a) 0.8801   (b) 0.1718

(c) 0.6491   (d) 0.5555

(e) 0.0923   (f) 0.0505

To the nearest hundredth:

(g) 0.9551   (h) 02192

(i) 0.0916   (j) 0.0774

(k) 0.9097   (l) 0.9999

To the nearest thousandth:

(m) 0.3138   (n) 0.09001

(o) 0.0091   (p) 0.4375

(q) 0.4426   (r) 0.0541

5. Convert the following common fractions to decimal fractions and round as indicated:

To the nearest hundredth:

(a) $\frac{2}{3}$ (b) $\frac{1}{4}$

(c) $\frac{1111}{10000}$ (d) $\frac{987}{1000}$

(e) $\frac{1001}{10000}$ (f) $\frac{21}{43}$

To the nearest thousandth:

(g) $\frac{15}{16}$ (h) $\frac{3}{7}$

(i) $\frac{1}{16}$ (j) $\frac{21}{43}$

(k) $\frac{7666}{10000}$ (l) $\frac{75196}{100000}$

To the nearest ten thousandth:

(m) $\frac{1}{64}$ (n) $\frac{11}{15}$

(o) $\frac{12345}{100000}$ (p) $\frac{1}{9}$

(q) $\frac{198}{205}$ (r) $\frac{31}{32}$

ADDITION AND SUBTRACTION OF DECIMAL FRACTIONS

INTRODUCTION

Addition and subtraction of decimal fractions are explained together in this unit because the two operations share important characteristics.

METHOD USED TO ADD DECIMAL FRACTIONS

To add decimal numbers, write them in a column with the decimal points lined up, as follows:

$$
\begin{array}{r}
7.2 \\
19.01 \\
.6 \\
+\,100.407 \\
\hline
127.217
\end{array}
$$

Be sure the decimal points are accurately lined up. You may find it helpful to add zeros so the right side of the column is filled:

$$
\begin{array}{r}
7.200 \\
19.010 \\
.600 \\
+\,100.407 \\
\hline
127.217
\end{array}
$$

METHOD USED TO SUBTRACT DECIMAL FRACTIONS

To subtract, you must also line up the decimal points.

$$
\begin{array}{r}
34.94 \\
-\,13.81 \\
\hline
21.13
\end{array}
$$

The operation is the same as that for whole numbers except for the decimal point. Review Unit Three on the Subtraction of Whole Numbers to brush up on carrying to fill place values. The decimal place values are treated just the same as the values to the left of the decimal point.

Unit 12—Practicing Addition and Subtraction of Decimal Fractions

Show all your work. Be certain the columns line up. Box your answers.

1. Add the following:

   (a) 9.3 + 0.2 + 1.6 =

   (b) 2.08 + 4.160 + 0.69 =

   (c) 345.009 + 0.345 + 9.034 =

2. Add the following:

   (a)   3.41   (b)   0.00532   (c)   0.109

            2.45   0.138   69.96

            4.67   4.322   10.01

       + 5.26   + 48.2   + 0.2

3. Add the following:

   (a) 16.601 + 0.195 + 4.749 + 0.945 + 200 =

   (b) 0.1741 + 0.049 + 33.0 + 0.1 + 2.3 =

   (c) 0.6593 + 0.4978 + 100.0 + 3.1416 + 0.0101 =

4. Subtract the following:

   (a)   0.9   (b)   19.51   (c)   60.0782
        − 0.2   − 4.19   − 42.38

5. Subtract the following:

   (a)   95.786   (b)   4.9   (c)   37.421
       − 88.999   − 0.807   − 37.980

6. Subtract the following:

   (a) $11.2 - 6.1356 =$

   (b) $29.006 - 8.0005 =$

   (c) $345.5842 - 0.095 =$

7. A pipe has a wall thickness of 0.126 inches. If the inside diameter is 3.617 inches, find the outside diameter.

8. A bar of cast iron 18.125 in. long has been tapered from a diameter of 0.983 to a diameter of 2.74 in. What is the difference in diameter between the two ends?

9. A company can produce a metal oil pan for a cost of $11.87. To make a profit of $4.98, at what price should it be sold?

10. Two pieces of metal measuring 65.283 in. and 23.014 in. long are welded together using a root opening of 0.031 in. What length is the final piece?

11. A salesperson estimated traveling expenses for May would be $500.00. At the end of May, the receipts were as follows: gasoline, $62.90; oil, $2.28; meals, $288.97; room, $285.83; miscellaneous, $18.46.
   (a) What were the total traveling expenses?
   (b) Were the expenses greater or less than the estimated amount and by how much?

12. A steel pipe 18.053 in. long has a 0.250 in. cap welded to each end. The caps are then machined so that 0.0625 in. of material are removed from each cap. What is the final length of the weldment?

13. Calculate the total of all the dimensions shown on this drawing.

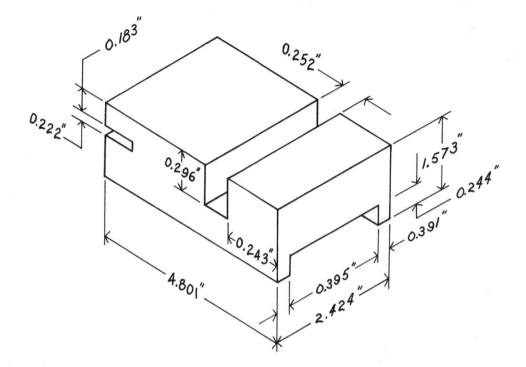

14. Each rod in this weldment is 0.5643 in. longer than the one before it. Rod Ⓐ is 0.7596 in. long.
    (a) What is the length of each of the remaining eight rods?
    (b) What is the total length of rod used in this weldment?

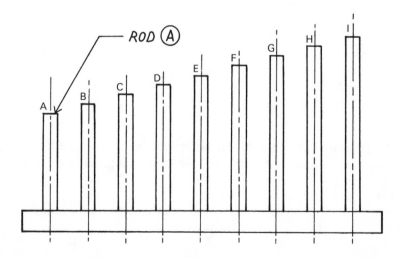

15. Calculate distance (A) .

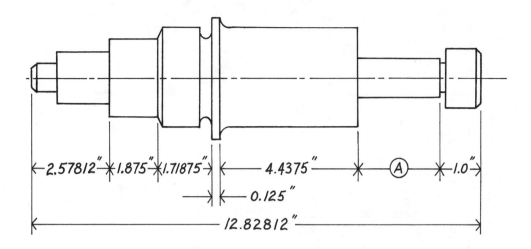

16. Grind 0.01625 in. from one side of each block and 0.02813 in. from the opposite side.
    (a) Calculate the finished thickness of each block.
    (b) Calculate the total finished thickness of all three blocks.

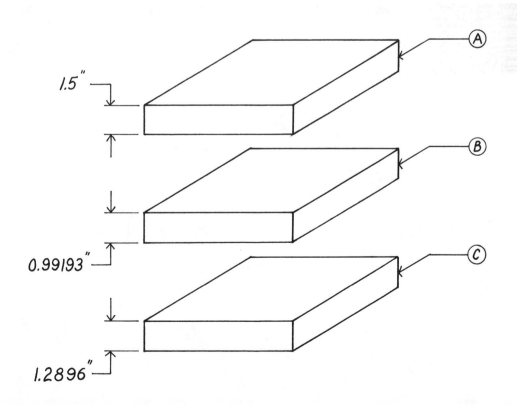

17. Calculate the total length of the six contact surfaces.

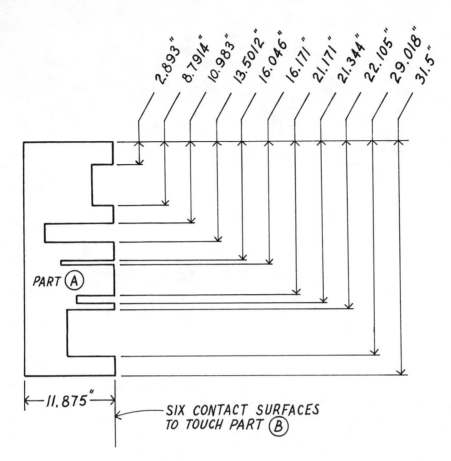

18. What are the overall dimensions of the block after 0.921 in. is machined from each surface?

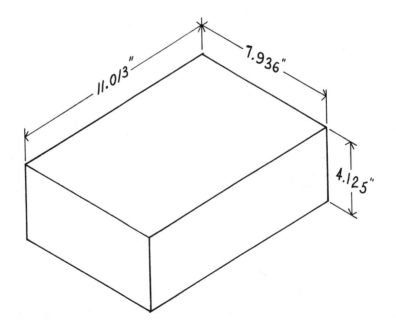

MULTIPLICATION OF DECIMAL FRACTIONS

INTRODUCTION

Multiplying decimal numbers is almost exactly the same as multiplying whole numbers. Refer back to Unit 4 if you need to refresh yourself on the topic of multiplication.

METHOD USED TO MULTIPLY DECIMAL FRACTIONS

When multiplying decimal numbers, the decimal points do not have to be lined up. Instead, you must line up the digits on the right side, as follows:

$$
\begin{array}{r}
14.124 \\
\times\,.31 \\
\hline
\end{array}
$$

Now, multiply in the same way you would for whole numbers. You can ignore the decimal point until you complete the multiplication process.

$$
\begin{array}{r}
14.124 \\
\times\,.31 \\
\hline
14124 \\
42372 \\
\hline
437844 \\
\end{array}
$$

When you have arrived at an answer, you will have to figure out where the decimal should be placed. Here's how. Count the total number of figures to the right of the decimal point in the two numbers being multiplied. In this example, the total is five. Now, count off five figures from the right in the answer and place the decimal in front of the fifth figure. The answer here is 4.37844.

Occasions arise where you do not have enough digits in the answer to locate the decimal point. For example:

$$
\begin{array}{r}
.0234 \\
\times\,.2 \\
\hline
0468 \\
\end{array}
$$

This problem contains a total of five figures to the right of the decimal, but the answer produces only four figures. Whenever this occurs, add as many zeros as needed to the left of the answer. The answer here is 0.00468.

A different situation is illustrated by the next example.

$$\begin{array}{r} .875 \\ \times 2.4 \\ \hline 3500 \\ 1750 \\ \hline 2.1000 \end{array}$$

When the answer ends in zeros, they may be eliminated. The answer here is 2.1.

The final example illustrates adding zeros to the left and removing zeros from the right of the answer.

$$\begin{array}{r} .00684 \\ \times .25 \\ \hline 03420 \\ 01368 \\ \hline 017100 \end{array}$$

The answer is 0.00171

MULTIPLICATION OF DECIMALS BY 10, 100, 1000, ETC.

When a decimal number is multiplied by 10, 100, 1000, 10000, etc., the answer can be quickly calculated. Move the decimal to the <u>right</u> the same number of places as there are zeros in the multiplier. These examples will illustrate:

$$\begin{array}{rclcl} 1.8 & \times & 10 & = & 18. \\ 1.8 & \times & 100 & = & 180. \\ 1.8 & \times & 1000 & = & 1800. \\ .645 & \times & 100 & = & 64.5 \\ 2,600,000 & \times & 10000 & = & 26,000,000,000 \end{array}$$

Unit 13—Practicing Multiplication of Decimal Fractions

Show all your work. Box your answers.

1. Multiply the following:

(a)
$$\begin{array}{r} .98 \\ \times .44 \end{array}$$

(b)
$$\begin{array}{r} .409 \\ \times 9.04 \end{array}$$

(c)
$$\begin{array}{r} .572 \\ \times 123 \end{array}$$

(d)
$$\begin{array}{r} 8.006 \\ \times .007 \end{array}$$

(e)
$$\begin{array}{r} 10.0001 \\ \times 100.1 \end{array}$$

(f)
$$\begin{array}{r} 875 \\ \times .25 \end{array}$$

2. Multiply the following:

   (a) 28 × .17 =

   (b) 18.125 × 29 =

   (c) 2500 × .4375 =

   (d) .632 × 22 =

   (e) 7.3 × 9.2 × .25 =

   (f) 6.6 × .99 × 11.0 =

3. Multiply the following:

   (a) 659.95 × .002 =

   (b) 7.9582 × 1.49 =

   (c) 100000 × 1.95 =

   (d) .0015 × 100 =

   (e) 2.423 × 8.97 × 3.0 =

   (f) 10.01 × 1000 × 1.0 =

4. What is the weight of 2,609 aluminum castings if each casting weighs 23.47 lbs.?

5. One cubic inch of steel weighs 0.2835 lbs. What is the weight of a block of steel containing 48.35 cubic inches? Round your answer to the nearest hundredth.

6. A salesperson offered you the following deal on some shop supplies: four hundred and sixty-five dollars in cash, or $80.00 down payment and 12 payments of $35.93 each. How much money would you save by paying cash?

7. What is the total cost to fill a storage tank with 5385.5 gallons of diesel fuel at 28.9 cents per gallon? Express your answer in dollars and cents to the nearest cent.

8. A pipe support is welded in 3.75 minutes and uses 4.3125 cubic feet of acetylene gas. How many cubic feet of gas are used to weld 12 of the supports?

9. An oxyacetylene cutting machine with 8 cutting torches uses 2.975 cubic feet of oxygen per torch to produce 8 cover blanks. How many cubic feet of oxygen are required to produce 8000 cover blanks?

10. Seven tube assemblies are required to complete a job. What is the total length of condenser tube required?

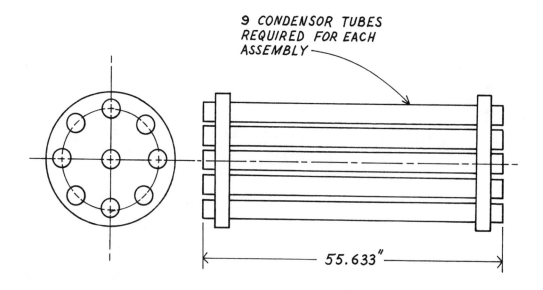

9 CONDENSOR TUBES REQUIRED FOR EACH ASSEMBLY

55.633"

11. Determine the total weight of the welded frames required for Job #8861.

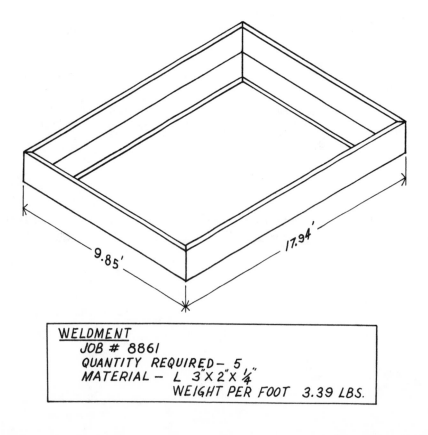

9.85'

17.94'

WELDMENT
JOB # 8861
QUANTITY REQUIRED— 5
MATERIAL — L 3"X 2"X ¼"
WEIGHT PER FOOT 3.39 LBS.

12. Calculate distance (A) .

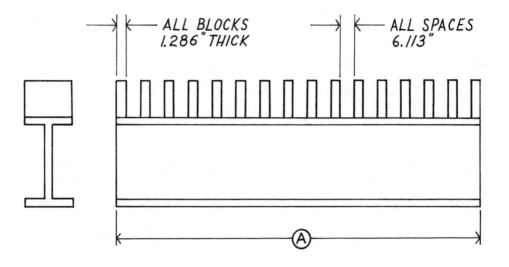

13. Twenty of the following braces are required. What is the total weight?

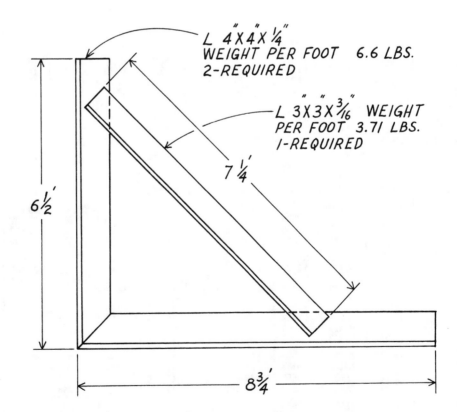

14. Should a truck designed for maximum load of 10,000 lbs. be used to deliver the following order?

> 20 pieces of 1 9/16 in. round bar at 12 ft.
> 500 pieces of 1/4 in. round bar at 9 ft.
> 70 pieces of 2 in. square bar at 6 ft.
> 5 pieces of 2 13/16 in. round bar at 8 ft.

SQUARE AND ROUND BARS
Weight and area

Size Inches	Weight Lb. per Foot ■	Weight Lb. per Foot ●	Area Square Inches ▨	Area Square Inches ◎	Size Inches	Weight Lb. per Foot ■	Weight Lb. per Foot ●	Area Square Inches ▨	Area Square Inches ◎
0	0.013	0.010	0.0039	0.0031	3	30.63	24.05	9.000	7.069
1/16	0.053	0.042	0.0156	0.0123	1/16	31.91	25.07	9.379	7.366
1/8	0.120	0.094	0.0352	0.0276	1/8	33.23	26.10	9.766	7.670
3/16					3/16	34.57	27.15	10.160	7.980
1/4	0.213	0.167	0.0625	0.0491	1/4	35.94	28.23	10.563	8.296
5/16	0.332	0.261	0.0977	0.0767	5/16	37.34	29.32	10.973	8.618
3/8	0.479	0.376	0.1406	0.1105	3/8	38.76	30.44	11.391	8.946
7/16	0.651	0.512	0.1914	0.1503	7/16	40.21	31.58	11.816	9.281
1/2	0.851	0.668	0.2500	0.1963	1/2	41.68	32.74	12.250	9.621
9/16	1.077	0.846	0.3164	0.2485	9/16	43.19	33.92	12.691	9.968
5/8	1.329	1.044	0.3906	0.3068	5/8	44.71	35.12	13.141	10.321
11/16	1.608	1.263	0.4727	0.3712	11/16	46.27	36.34	13.598	10.680
3/4	1.914	1.503	0.5625	0.4418	3/4	47.85	37.58	14.063	11.045
13/16	2.246	1.764	0.6602	0.5185	13/16	49.46	38.85	14.535	11.416
7/8	2.605	2.046	0.7656	0.6013	7/8	51.09	40.13	15.016	11.793
15/16	2.991	2.349	0.8789	0.6903	15/16	52.76	41.43	15.504	12.177
1	3.403	2.673	1.0000	0.7854	4	54.44	42.76	16.000	12.566
1/16	3.841	3.017	1.1289	0.8866	1/16	56.16	44.11	16.504	12.962
1/8	4.307	3.382	1.2656	0.9940	1/8	57.90	45.47	17.016	13.364
3/16	4.798	3.769	1.4102	1.1075	3/16	59.67	46.86	17.535	13.772
1/4	5.317	4.176	1.5625	1.2272	1/4	61.46	48.27	18.063	14.186
5/16	5.862	4.604	1.7227	1.3530	5/16	63.28	49.70	18.598	14.607
3/8	6.433	5.053	1.8906	1.4849	3/8	65.13	51.15	19.141	15.033
7/16	7.032	5.523	2.0664	1.6230	7/16	67.01	52.63	19.691	15.466
1/2	7.656	6.013	2.2500	1.7671	1/2	68.91	54.12	20.250	15.904
9/16	8.308	6.525	2.4414	1.9175	9/16	70.83	55.63	20.816	16.349
5/8	8.985	7.057	2.6406	2.0739	5/8	72.79	57.17	21.391	16.800
11/16	9.690	7.610	2.8477	2.2365	11/16	74.77	58.72	21.973	17.257
3/4	10.421	8.185	3.0625	2.4053	3/4	76.78	60.30	22.563	17.721
13/16	11.179	8.780	3.2852	2.5802	13/16	78.81	61.90	23.160	18.190
7/8	11.963	9.396	3.5156	2.7612	7/8	80.87	63.51	23.766	18.665
15/16	12.774	10.032	3.7539	2.9483	15/16	82.96	65.15	24.379	19.147
2	13.611	10.690	4.0000	3.1416	5	85.07	66.81	25.000	19.635
1/16	14.475	11.369	4.2539	3.3410	1/16	87.21	68.49	25.629	20.129
1/8	15.366	12.068	4.5156	3.5466	1/8	89.38	70.20	26.266	20.629
3/16	16.283	12.788	4.7852	3.7583	3/16	91.57	71.92	26.910	21.135
1/4	17.227	13.530	5.0625	3.9761	1/4	93.79	73.66	27.563	21.648
5/16	18.197	14.292	5.3477	4.2000	5/16	96.04	75.43	28.223	22.166
3/8	19.194	15.075	5.6406	4.4301	3/8	98.31	77.21	28.891	22.691
7/16	20.217	15.879	5.9414	4.6664	7/16	100.61	79.02	29.566	23.221
1/2	21.267	16.703	6.2500	4.9087	1/2	102.93	80.84	30.250	23.758
9/16	22.344	17.549	6.5664	5.1572	9/16	105.29	82.69	30.941	24.301
5/8	23.447	18.415	6.8906	5.4119	5/8	107.67	84.56	31.641	24.850
11/16	24.577	19.303	7.2227	5.6727	11/16	110.07	86.45	32.348	25.406
3/4	25.734	20.211	7.5625	5.9396	3/4	112.50	88.36	33.063	25.967
13/16	26.917	21.140	7.9102	6.2126	13/16	114.96	90.29	33.785	26.535
7/8	28.126	22.090	8.2656	6.4918	7/8	117.45	92.24	34.516	27.109
15/16	29.362	23.061	8.6289	6.7771	15/16	119.96	94.22	35.254	27.688
3	30.625	24.053	9.0000	7.0686	6	122.50	96.21	36.000	28.274

AMERICAN INSTITUTE OF STEEL CONSTRUCTION

15. What is the total weight of Job #8853?

QUANTITY	PART DESCRIPTION	WEIGHT
37	¾" STD. PIPE AT 6.25'	1.13 LBS. PER FT.
148	1" ROUND BAR AT 0.5685'	2.67 LBS. PER FT.
4	M SHAPE AT 10.5'	18.9 LBS. PER FT.
7	L 5"X3"X¼" AT 7.0625'	6.6 LBS. PER FT.
8	3" STD. CHANNEL AT 8.125"	4.1 LBS. PER FT.

PARTS LIST
JOB — 8853
DATE — AUGUST 29, 1988

DIVISION OF DECIMAL FRACTIONS

INTRODUCTION

Dividing decimal numbers is almost exactly the same as dividing whole numbers. Refer back to Unit 5 if you need to refresh yourself on the topic of division.

METHOD USED TO DIVIDE DECIMAL FRACTIONS

When dividing decimal numbers, your first concern is the decimal point. Here is what you must do. If there is a decimal point in the divisor, simply move it to the right of the last digit. You want to change the divisor to a whole number. Then move the decimal point in the dividend the same number of places to the right. Here's an example:

Divide 186.942 by 17.34 to the nearest thousandth:

$$17.34 \overline{)186.942}$$

Move the decimal in the divisor (17.34) two places to the right and then move the decimal in the dividend (186.942) two places to the right.

$$1734. \overline{)18694.2}$$

This changes the divisor to a whole number and greatly simplifies the task of division. Next, place a decimal directly above the decimal in the dividend. This is done so the decimal in the answer is properly located. Now, proceed with the division. Remember to calculate to the ten thousandth so the answer can be rounded to the nearest thousandth.

$$
\begin{array}{r}
10.7809 \\
1734. \overline{)18694.2000} \\
1734 \\
\hline
13542 \\
12138 \\
\hline
14040 \\
13872 \\
\hline
16800 \\
15606 \\
\hline
1194
\end{array}
$$

The answer is 10.781.

There is one more item you will need to know. The dividend may not have enough zeros to accommodate the relocated decimal. If this occurs, add enough zeros so you can locate the decimal properly.

$$9.345\overline{)27.25}$$

The decimal is relocated like this:

$$9345.\overline{)27250.}$$

DIVISION OF DECIMALS BY 10, 100, 1000, ETC.

When a decimal number is divided by 10, 100, 1000, 10000, etc., the answer can be quickly calculated. Move the decimal to the <u>left</u> the same number of places as there are zeros in the divisor. These examples will illustrate:

185	÷	10	=	18.5
185	÷	100	=	1.85
185	÷	1000	=	.185
96.88	÷	10000	=	.09688
7,500,000	÷	1,000,000	=	7.5

Unit 14—Practicing Division of Decimal Fractions

Show all your work. Box your answers.

1. Divide the following:

(a) 875 ÷ 0.25 =

(b) 1.17 ÷ 0.003 =

(c) 3.94 ÷ 20 =

(d) 10.001 ÷ 0.01 =

(e) 0.072 ÷ 0.009 =

(f) 6.8752 ÷ 1000 =

2. Divide the following to the nearest hundredth:

(a) $\dfrac{0.1217}{1.72} =$

(b) $\dfrac{7.58}{2.16} =$

(c) $\dfrac{100.449}{100} =$

(d) $\dfrac{96603}{100000} =$

(e) $\dfrac{6.3}{0.046} =$

(f) $\dfrac{71}{0.72} =$

3. Divide the following to the nearest thousandth:

(a) $16.5 \div 5.5 =$

(b) $48 \div 10.1 =$

(c) $17.2189 \div 1000 =$

(d) $6.9 \div 0.96 =$

(e) $6.9 \div 9.6 =$

(f) $2009.15 \div 100 =$

4. A gear pump can deliver 114.5 gallons of water per hour. How many hours would it take to empty a tank containing 1,822 gallons? Express your answer to the nearest hour.

5. One carton containing 9 one-gallon cans of cold galvanizing compound cost $127.80. How many one-gallon cans are you able to buy for $4,430.40?

6. A large container of stainless steel ball bearings weighs 212.55 lbs. Each ball bearing weighs 0.2834 lbs. How many ball bearings are in the container?

7. A surface grinder removes 0.013 in. of steel with each pass. How many passes are required to reduce a piece of steel from 2.095 in. to 1.744 in.?

8. How many precision cut blocks of steel 0.869 in. thick will fit into an opening of 3.652 in.? How much space, if any, will remain in the opening after the blocks have been inserted?

9. A company offered the following deal on a standard item they produced. The first 150 items a customer ordered would cost $5,227.50. Each item after that would cost the customer $30.50. If the company received a $19,654.00 order, how many items were in the order?

10. Divide this channel into 11 equal pieces. What is the length of each piece?

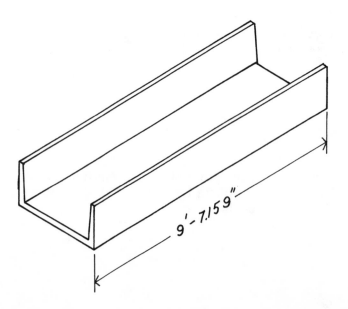

9' - 7.159"

11. How many 1.857 in. squares can be cut from this sheet of steel?

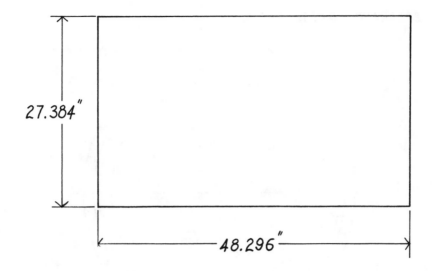

12. This square bar weighs 156.686 lbs. What is the weight of a 1 in. piece?

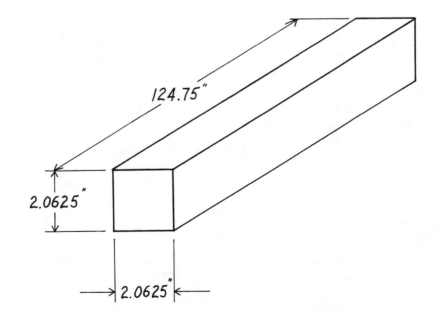

13. The clearance between the collar and drum shown below is 0.0573 in. Using a thermal coating process, the clearance is to be reduced to 0.0087 in. Each coating application deposits 0.0009 in. of material. How many applications are required?

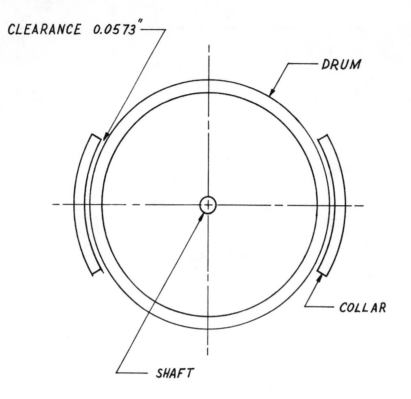

14. How many of the following drums would there be in a stack 47.3664 in. high?

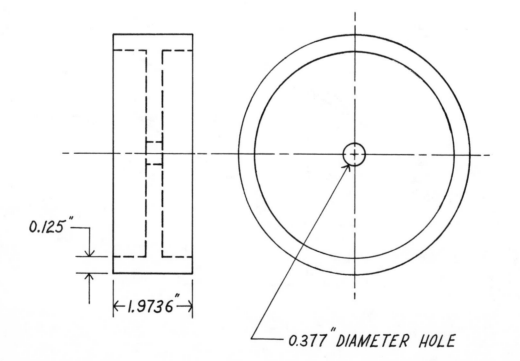

15. What is the distance between the evenly spaced vertical blocks?

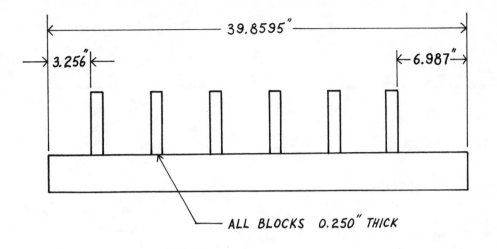

Section Four

MEASUREMENT

Objectives for Section Four, Measurement

After studying this section, you will be able to:

- Define linear measure, angular measure, and circular measure
- Show how to convert linear measurements between units
- Explain and give examples of various forms of tolerance
- Show how to perform angular measurement
- Perform four basic operations for angular calculations
- Define basic four-sided shapes
- Calculate perimeter and area for four-sided shapes
- Convert square units of measure
- Define basic triangular shapes
- Calculate perimeter and area for triangular shapes
- Define various parts of circular shapes
- Calculate circumference and area of circular shapes

Contents for Section Four, Measurement

LINEAR MEASURE

INTRODUCTION

A welder's job includes taking and reading measurements. Generally, as a welder, you will be dealing with the three types of measurements listed below:

1. **Linear measure** refers to measuring the straight line distance between two points.
2. **Angular measure** refers to measuring the angle formed by two intersecting lines.
3. **Circular measure** refers to measuring curved lines.

LINEAR MEASURE

Two measuring systems are presently in use by welders, the customary and the metric. You should become very familiar with both systems.

CUSTOMARY SYSTEM

Customary system is the most common measuring system we are all familiar with. The basic unit is the inch. Twelve inches make up one foot and 36 inches form one yard. In welding, you will be concerned only with feet and inches. The standard notation, for example, is 6 ft. 11 in.

METRIC SYSTEM

You are taking your training just at the time the welding industry, along with the rest of industrial America, is changing from the customary system to the **metric system.** A full explanation of metrics as it relates to your work is provided in Unit 24. Following is a short explanation of metric line measurement only.

THE MILLIMETER

The smallest unit you will encounter in the shop or in the field is the **millimeter.** In fact, it is probably the only metric unit you will use because almost all metric blueprints are dimensioned only in millimeters. It is the basic unit in metric dimensioning, just as the inch is the basic unit in the customary system. An example of the notation used for millimeters is **mm.** As you can see in the illustration below, it is a very small unit of measurement.

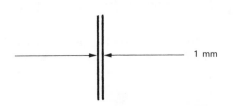

Here are two ways to visualize a millimeter.

1. A millimeter is approximately the thickness of a line made with a dull pencil.
2. A dime is about one millimeter thick.

Since millimeters are very small, they accumulate very quickly. For example, there are 25.4 millimeters in one inch. If you are 5 ft. 10 in. tall, then you are 1 778 mm tall. A 10 ft. bar is 3 048 mm long. Eventually, through practice, you will develop a "feel" for metric lengths. If you do not own a metric scale, you should acquire one now. Note that in the metric system, a comma is not placed between sets of three values.

CONVERTING LINEAR MEASUREMENTS

Your shop work will require you to change back and forth between inches, feet, and millimeters. Often, these conversions will result in lengthy decimals, so it is common practice to round off conversions. Also, you will sometimes have to convert decimal dimensions to fractional dimensions normally found on your scale (eighths, sixteenths, etc.).

CONVERTING FEET TO INCHES

To convert from feet to inches, multiply the number of feet by twelve.

$$7 \text{ ft.} = 7 \times 12 = 84 \text{ in.}$$

$$27.35 \text{ ft.} = 27.35 \times 12 = 328.2 \text{ in.}$$

$$4\frac{19}{64} \text{ ft.} = \frac{275}{64} \times 12 = 51.5625 = 51\frac{9}{16} \text{ in.}$$

CONVERTING INCHES TO FEET

To convert from inches to feet, divide the number of inches by twelve. The answer is in its most useful form when expressed in feet and inches rather than only in feet, as the following examples will illustrate.

$$109 \text{ in.} = 109 \div 12 = 9 \text{ ft. } 1 \text{ in.}$$

$$219\frac{1}{4} \text{ in.} = 219\frac{1}{4} \div 12 = 18 \text{ ft. } 3\frac{1}{4} \text{ in.}$$

$$28.7 \text{ in.} = 28.7 \div 12 = 2 \text{ ft. } 4.7 \text{ in.}$$

CONVERTING INCHES TO MILLIMETERS

To convert from inches to millimeters, multiply the number of inches by 25.4. A good rule of thumb for rounding the answer is to express it one <u>less</u> decimal place than the question.

$$16 \text{ in.} = 16 \times 25.4 = 406.4 = 406 \text{ mm}$$

$$93.375 \text{ in.} = 93.375 \times 25.4 = 2371.725 = 2371.73 \text{ mm}$$

$$\frac{1}{16} = \frac{1}{16} \times 25.4 = 1.5875 = 1.6 \text{ mm}$$

CONVERTING MILLIMETERS TO INCHES

To convert from millimeters to inches, divide the number of inches by 25.4. A good rule of thumb for rounding the answer is to express it one <u>more</u> decimal place than the question.

$$1\ 789 \text{ mm} = 1789 \div 25.4 = 70.4 \text{ in.}$$

$$9.32 \text{ mm} = 9.32 \div 25.4 = .3669 = .367 \text{ in.}$$

$$360.5 \text{ mm} = 360.5 \div 25.4 = 14.19 \text{ in.}$$

CONVERTING DECIMALS TO FRACTIONS

In Unit 11 you learned to convert decimals to fractions. A new situation arises when doing such a conversion for the purpose of linear measure. Here, you will have to convert the decimal dimension to one of the fractional dimensions normally found on your steel rule, such as eighths, sixteenths, thirty-seconds, or sixty-fourths. A typical example would be to convert 10.21 in. to a fractional dimension to the nearest sixty-fourth.

1. Dealing with the decimal part only, 0.21 converts to $\frac{21}{100}$.

2. Since 21/100 will not reduce exactly to a fraction with a denominator of 64, it will have to be reduced to the nearest sixty-fourth.

3. Begin this process by writing an equation like this:

$$\frac{21}{100} = \frac{x}{64}$$

Here's what this means. Since you want the final fraction to be in sixty-fourths, you know the denominator will be 64. However, you do not yet know what the numerator will be, so it is identified for now as "x."

4. The method for finding x can be expressed in one phrase: <u>Cross multiply and divide.</u>

   1. $\frac{21}{100} \searrow \frac{x}{64}$

   2. $21 \times 64 = 1344$

   3. $1344 \div 100 = 13\frac{11}{25}$

   Remember, you are trying to find the number that will replace the x in the fraction x/64. That number as calculated so far would be 13 11/25. But, instead of using 13 11/25, round to the nearest whole number, which is 13. Thirteen, then, is the numerator.

5. The answer is $10\frac{13}{64}$ in. (to the nearest sixty-fourth).

SUMMARY OF CONVERSION FACTORS		
FROM	**TO**	**METHOD**
Feet	Inches	Multiply by 12
Inches	Feet	Divide by 12
Inches	Millimeters	Multiply by 25.4
Millimeters	Inches	Divide by 25.4

TOLERANCE

Taking accurate measurements is obviously important to any tradesperson. But there is a limit to how accurately you can measure. This limit is determined by two main factors; the accuracy of the measuring tool and the accuracy of the surface of the material being measured. Because of this, dimensions on blueprints normally indicate a certain acceptable range that you are allowed to work within. This range is called the **tolerance** and it is usually indicated as shown in the following examples:

TOLERANCE + 1/8 in.

This indicates an object may exceed the given dimension by 1/8 in. Given a tolerance of + 1/8 in., an object dimensioned 48 3/8 in. long would have a maximum acceptable dimension of 48 1/2 in. and a minimum of 48 3/8 in.

TOLERANCE − 0.0625 in.

This indicates an object may be 0.0625 in. less than the given dimension. Given a tolerance of 0.0625 in., an object dimensioned 99.6 in. long would have a maximum acceptable dimension in. and a minimum of 48 3/8 in.

TOLERANCE ± 3 mm

This indicates an object may be 3 mm greater or less than the given dimension. In this case, you have a range of 6 mm to work within. Given a tolerance of ± 3 mm, an object dimensioned 211 mm long would have a maximum acceptable dimension of 214 mm and a minimum of 208 mm.

KEY TERMS FOR WELDERS

LINEAR MEASURE
ANGULAR MEASURE
CIRCULAR MEASURE
CUSTOMARY SYSTEM
METRIC SYSTEM
MILLIMETER
mm
TOLERANCE
±

Unit 15—Practicing Linear Measurements

Show all your work. Box your answers.

1. Measure the following lines as accurately as possible, in millimeters and inches.

(a) _____

(b) ____

(c) _____

(d) _____

(e) _____

(f) _____

2. Measure the following as accurately as possible in millimeters and inches.

   (a) Your height.

   (b) The height of a coffee cup.

   (c) The length of a welding rod.

   (d) The height of a workbench or desk.

   (e) The span of your hand from smallest finger to thumb.

   (f) The width of a doorway.

3. Convert the following feet to inches.

   (a) 35 − 1/4 ft.   (b) 123 ft.

   (c) 45.375 ft.   (d) 17 5/23 ft.

   (e) 503.5 ft.   (f) 4 1/8 ft.

   (g) 5,280 ft.   (h) 65.1 ft.

   (i) 17/64 ft.   (j) 29.47 ft.

   (k) 12 1/12 ft.   (l) 15/32 ft.

4. Convert the following inches to feet (and inches if necessary).

   (a) 1 1/2 in.   (b) 6,000 in.

   (c) 95.75 in.   (d) 2/3 in.

   (e) 215 1/32 in.   (f) 12 in.

   (g) 787.96 in.   (h) 17/64 in.

   (i) 83,917 in.   (j) 38 1/4 in.

   (k) 4 1/8 in.   (l) 0.29 in.

5. Convert the following inches to millimeters (to the nearest tenth of a millimeter).

   (a) 12 in.

   (b) 36 in.

   (c) 10 in.

   (d) 1 in.

   (e) 1/64 in.

   (f) 69 in.

   (g) 5/8 in.

   (h) 1,000 in.

   (i) 63 3/16 in.

   (j) 59 in.

   (k) 8 1/4 in.

   (l) 480 in.

6. Convert the following millimeters to inches (to the nearest sixty-fourth).

   (a) 1 000 mm

   (b) 25.4 mm

   (c) 1.0 mm

   (d) 279.4 mm

   (e) 65.5 mm

   (f) 12 mm

   (g) 11 850 mm

   (h) 35 600 mm

   (i) 195 mm

   (j) 862.3 mm

   (k) 5 280 mm

   (l) 100 mm

7. Convert the following decimal inches to fractional dimensions.
   Convert to the nearest eighth of an inch.

   (a) 15.8 in.

   (b) 0.982 in.

   (c) 48.012 in.

   (d) 80.125 in.

   (e) 19.375 in.

   (f) 0.9126 in.

   Convert to the nearest sixteenth of an inch.

   (g) 19. 85 in.

   (h) 32.032 in.

   (i) 0.997 in.

   (j) 855.5 in.

   (k) 21.0625 in.

   (l) 1010.7 in.

Convert to the nearest thirty-second of an inch.

(m) 11.65 in.   (n) 58.022 in.

(o) 730.18 in.   (p) 0.28125 in.

(q) 0.979 in.   (r) 3602.79 in.

Convert to the nearest sixty-fourth of an inch.

(s) 68.033 in.   (t) 0.908 in.

(u) 12.55 in.   (v) 0.578125 in.

(w) 632.41 in.)   (x) 7802.77 in.

8. What are the maximum and minimum dimensions allowed in the following lengths?

	Dimension	Tolerance	Max	Min
(a)	249 mm	− 2 mm		
(b)	17 1/32 in.	+ 1/16 in.		
(c)	1.340 in.	± 0.005 in.		
(d)	8 ft.-11 3/4 in.	+ 1/2 in.		
(e)	14.5 mm	± 0.5 mm		
(f)	18 9/16 in.	+ 1/32 in.		
(g)	19.9 mm	± 0.2 mm		
(h)	8 ft.-0 in.	− 1/16 in.		

9. Convert the following dimensions to fractional dimensions (to the nearest one-hundredth of an inch). Then, using a tolerance of ± 3/100 in., calculate the maximum and minimum allowable dimensions.

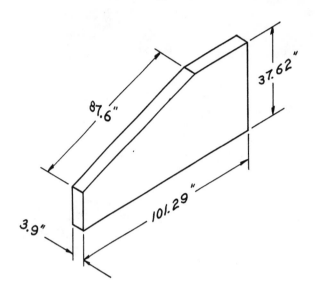

10. The tolerance for the hole in this weldment is + 1/8 in. All other dimensions have a tolerance of ± 1/16 in. Prepare a list of all dimensions from smallest to largest and the maximum and minimum dimensions allowed.

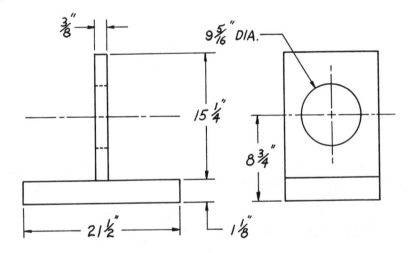

11. Prepare the following lists from the drawing below.
    (a) A list of the dimensions from smallest to largest.
    (b) A list of the decimal equivalents to three decimal places.
    (c) A list of the metric equivalents in millimeters to two decimal places.

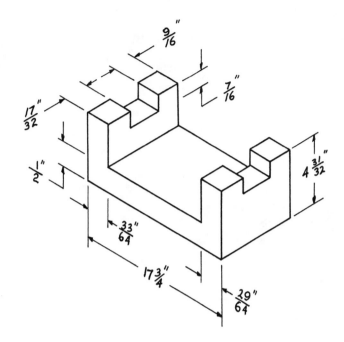

12. Given a tolerance of − .5 mm, prepare a list, from smallest to largest, of the given dimensions with the maximum and minimum allowable dimensions.

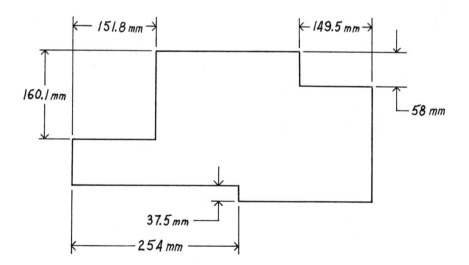

ANGULAR MEASURE

INTRODUCTION

An **angle** can be defined as the opening between two intersecting lines.

Welders are often called upon to work with angles. Therefore, you should be able to take measurements of angles and to perform math operations involving angles.

ANGULAR MEASUREMENT

The units of measure used for angles are **degrees, minutes,** and **seconds.** The largest unit is the degree. Mathematicians have divided the circle into 360 parts, calling each part a degree. Therefore, 1 degree is 1/360 of a circle. The notation for degrees is °. To provide a more accurate measurement, each degree has been divided into 60 equal parts, called minutes, and noted as ′. To allow for even more accurate measurement, each minute has been subdivided into 60 equal parts called seconds, and noted as ″. (Unfortunately, the notation for minutes and seconds is the same as that used for feet and inches in linear measure.)

Angles can be measured using an instrument called a **protractor,** as shown below.

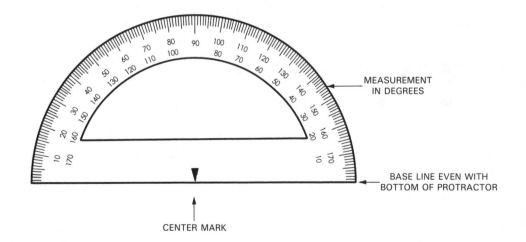

To use a protractor, the base of the protractor should align with one side of the angle. Place the center marker at the intersection of the lines making the angle. Then, position the protractor as shown and read the angle as 55°.

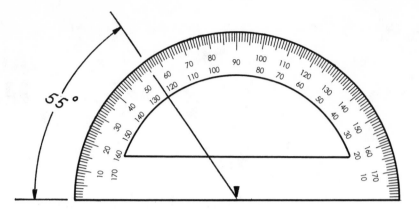

Since the degree of accuracy in measuring angles is limited by the tools being used and the material being measured, your shop work is normally allowed a tolerance. As with linear measure, the tolerance may be +, −, or ±. A typical example on a blueprint may call for an angle of 53° ±2°. In this case, the angle could vary from 55° to 51° and still be acceptable.

ANGULAR CALCULATION

The basic math operations of addition, subtraction, multiplication, and division can be performed with angles.

ADDITION OF ANGLES

When adding, line up the units of measure (degrees, minutes, seconds) as shown and add each unit.

$$
\begin{array}{rrr}
45° & 39' & 17'' \\
+18° & 55' & 40'' \\
\hline
63° & 94' & 57''
\end{array}
$$

In many cases, the answer will produce minutes or seconds of 60 or more. In the above example, the minutes total 94. Since 60′ are equal to one degree, add one degree to the total degrees and remove 60' from the total minutes. The final answer is 64° 34' 57". If the seconds were 60 or greater, one minute would be added to the minutes and 60 seconds removed from the seconds.

SUBTRACTION OF ANGLES

Line up the units as in addition.

$$
\begin{array}{rrr}
11° & 52' & 10'' \\
- 7° & 29' & 37'' \\
\hline
\end{array}
$$

In this example, 37″ cannot be subtracted from 10″. Borrow 1′ from the 52′, add 60″ to the 10″, and then subtract.

$$
\begin{array}{rrr}
11° & 51' & 70'' \\
- 7° & 29' & 37'' \\
\hline
4° & 22' & 33''
\end{array}
$$

MULTIPLICATION OF ANGLES

Multiply each of the degrees, minutes, and seconds as shown.

$$
\begin{array}{rrr}
30° & 47' & 22'' \\
 & & \times\ 4 \\
\hline
120° & 188' & 88''
\end{array}
$$

Reduce the 88″ by 60″ and add 1′ to 188′ to arrive at 189′. Reduce the 189′ by 180′ (3 degrees) and increase the 120° to 123°. The final answer is 123° 9′ 28″.

DIVISION OF ANGLES

Begin by dividing the degrees. If a remainder occurs, change it to minutes and add it to the existing minutes. Now, divide the total number of minutes. Again, if a remainder occurs, add it to the seconds. Next, divide the total number of seconds. If a remainder occurs, round it to the nearest second.

```
            40°   34'    44"
    3 /121°  44'    14"
      12
      ─────
      01 →  60
            104
              9
            ─────
             14
             12
              2 → 120
                  134
                   12
                  ─────
                   14
                   12
                    2
```

The answer is 40° 34′ 45″ rounded to the nearest second.

KEY TERMS FOR WELDERS
ANGLE
DEGREE
MINUTE
SECOND
PROTRACTOR

Unit 16—Practicing Angular Measurement

Show all your work. Be certain the columns line up. Box your answers.

1. Using a protractor, measure the following angles as accurately as possible.

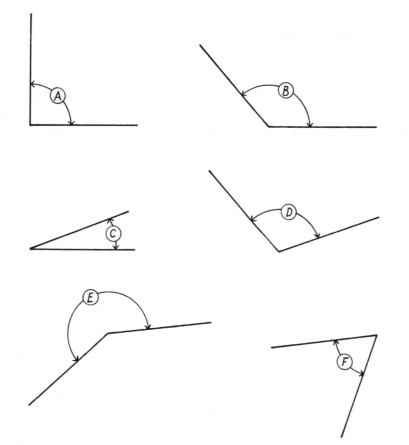

2. Add the following angles and draw the resulting angle, as accurately as possible, using a protractor.

(a) $7°$ $20'$
    $+\ 3°$ $18'$

(b) $92°$ $36'$ $10''$
    $+$ $13'$ $35''$

(c) $45°$ $45'$ $9''$
    $+29°$ $35'$ $51''$

(d) $152°$ $32'$ $54''$
    $+27°$ $27'$ $6''$

(e)   109° 51′ 44″
  + 90° 47′ 39″

(f)   91° 55′ 12″
  +76°   2′ 14″

(g)   55° 13′ 16″
  +11° 37′ 39″

(h)   19° 50′   1″
     4°   6′ 19″
  +32°   9′ 20″

3. Subtract the following angles and draw the resulting angle, as accurately as possible, using a protractor.

(a)   119° 17′ 43″
  − 91° 15′ 23″

(b)   10° 40′
  − 7° 37′

(c)   135° 54′ 11″
  −101° 55′   9″

(d)   204°   6′ 32″
  −69°   3′ 42″

(e)   652° 44′   0″
  −359°   0′ 18″

(f)   80° 22′ 31″
  −79° 31′ 59″

(g)   45°   0′ 21″
  −33°   7′ 53″

(h)   86° 36′ 12″
  −   23′ 14″

4. Multiply the following angles:

(a) 35° 24′ 22″
     × 8

(b) 46° 11′ 48″
     × 9

(c) 129° 58′ 36″
     × 7

(d) 90° 43′
     × 5

(e) 206° 32′ 47″
     × 6

(f) 18° 14′ 3″
     × 12

5. Divide the following angles and draw the resulting angle, as accurately as possible, using a protractor.

(a) 3 / 93° 42′ 27″

(b) 2 / 181° 13′ 50″

(c) 6 / 180°

(d) 8 / 222°

(e) 7 / 360°

(f) 8 / 360°

(g) 11 / 93° 7′ 45″

(h) 27 / 19° 14′ 57″

6. Calculate the maximum and minimum angles according to the given tolerance.

	Angle	Tolerance	Maximum	Minimum
(a)	90°	− 5°		
(b)	101°	+ 30′		
(c)	37°	± 3° 35′		
(d)	83° 30′	± 50′		
(e)	30°	± 30″		
(f)	45°	+ 1° 15′		
(g)	212° 47′	− 30′		
(h)	66°	± 20′		

7. Using a protractor, draw angles of the following sizes:

(a) 90°   (b) 28°

(c) 261°   (d) 180°

(e) 125°   (f) 350°

(g) 15°   (h) 110°

8. Twelve studs are to be welded in a circle. What is the angle between each stud? Sketch the design.

9. Sixteen holes are to be drilled in a round flange. What is the angle between each hole? Sketch your answer.

10. What size is each angle when 180° is divided into 9 equal parts?

FOUR-SIDED FIGURE MEASURE

INTRODUCTION

The products that you will work on in the shop or at the job site are made up of regular geometric shapes that can be measured. Products such as frames, braces, drums, bins, hoppers, chutes, and tanks are examples of such items. In this unit you will learn to calculate the perimeter and area of the most common four-sided shapes found in industry. The **perimeter** is the distance around a shape. The **area** is the surface measure of a shape.

SQUARE

A **square** is a shape having four sides, the opposite sides are parallel, all sides are the same length, and all angles in the square are 90°.

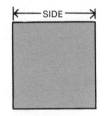

PERIMETER OF A SQUARE

To calculate the perimeter of a square, add the four sides.

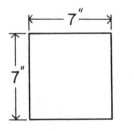

$$P = 7 + 7 + 7 + 7$$
$$P = 28''$$

AREA OF A SQUARE

To calculate the area of a square, multiply the two sides.

$$A = 7 \times 7$$
$$A = 49 \text{ square inches}$$

Area is always expressed in square units such as square feet, square millimeters, square miles, etc. Common notations for area include:

> 49 square feet
> 49 sq/ft
> 49 ft²
> 49 sq. ft.
> 750 mm²

RECTANGLE

A **rectangle** is a shape having four sides, the opposite sides are parallel and equal in length, and all the angles are 90°. It is similar to a square except one side is longer (length) than the adjacent side (width).

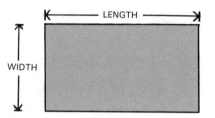

PERIMETER OF A RECTANGLE

To calculate the perimeter of a rectangle, add the four sides.

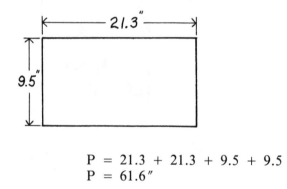

> P = 21.3 + 21.3 + 9.5 + 9.5
> P = 61.6″

AREA OF A RECTANGLE

To calculate the area of a rectangle, multiply length by width.

> A = 21.3 × 9.5
> A = 202.35 sq. in.

PARALLELOGRAM

A **parallelogram** has four sides and the sides opposite each other are the same length and are parallel. The angles opposite each other are equal. It looks like a rectangle that has been tilted.

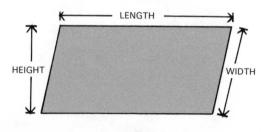

PERIMETER OF A PARALLELOGRAM

To calculate the perimeter of a parallelogram, add the four sides.

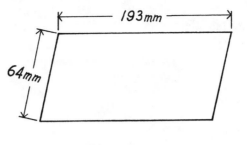

P = 193 + 193 + 64 + 64
P = 514 mm

AREA OF A PARALLELOGRAM

To calculate the area of a parallelogram, multiply the length by the height.

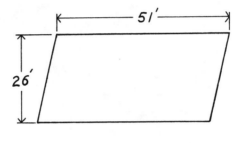

A = 51 × 26
A = 1326 sq. ft.

Notice it is <u>not</u> the lengths of the sides that are being multiplied, but the length of one side and the height. You may find it interesting to note that geometric shapes are often closely related. If you were to cut the parallelogram below along the dotted line and then reassemble it as shown, you would have created a rectangle.

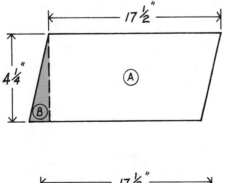

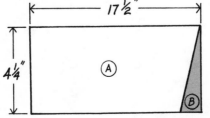

TRAPEZOID

The **trapezoid,** like all the shapes studied so far, has four sides. Two of the sides are parallel, but the other two sides are not parallel to each other. For this reason, a trapezoid is not as symmetric or pleasing to look at as a parallelogram. Also, it does not appear in fabrication and construction as often as the parallelogram.

PERIMETER OF A TRAPEZOID

To calculate the perimeter of a trapezoid, add the four sides.

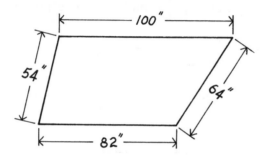

P = 54 + 100 + 64 + 82
P = 300″

AREA OF A TRAPEZOID

To calculate the area of a trapezoid:
1. Add the two parallel lengths.
2. Multiply by the height.
3. Divide by 2.

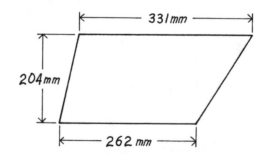

Example:
1. 331 + 262 = 593
2. 593 × 204 = 120,972
3. 120,972 ÷ 2 = 60,486 mm

As stated previously, geometric shapes are often related. If you make two trapeziums of the same size and then flip one over and butt them end to end, you will have formed a parallelogram which is twice the size of the original trapezoid.

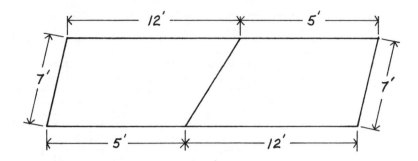

CONVERTING SQUARE UNITS OF MEASURE

In the welding trade, the most common units of square measure are square feet, square inches, and square millimeters. Each unit of measure can be converted to the other units.

SQUARE FEET TO SQUARE INCHES

A square with sides measuring one foot has an area of one square foot. Since one foot equals 12 inches, the square also has an area of 12 in. × 12 in. or 144 sq. in.

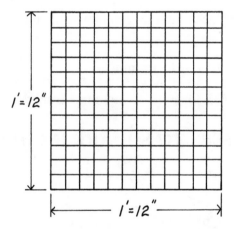

Therefore, to change square feet to square inches, multiply the number of square feet by 144.

$$1 \text{ sq. ft.} = 1 \times 144 = 144 \text{ sq. in.}$$

SQUARE INCHES TO SQUARE FEET

To convert square inches to square feet, divide the number of square inches by 144.

1. 144 sq. in. = 144 ÷ 144 = 1 sq. ft.
2. 1000 sq. in. = 1000 ÷ 144 = 6.9 sq. ft. (rounded)
3. 3600 sq. in. = 3600 ÷ 144 = 25 sq. ft.

SQUARE INCHES TO SQUARE MILLIMETERS

A square with sides measuring one inch has an area of one square inch. Since one inch equals 25.4 mm, the square also has an area of 25.4 in. × 25.4 in. or 645.16 square millimeters.

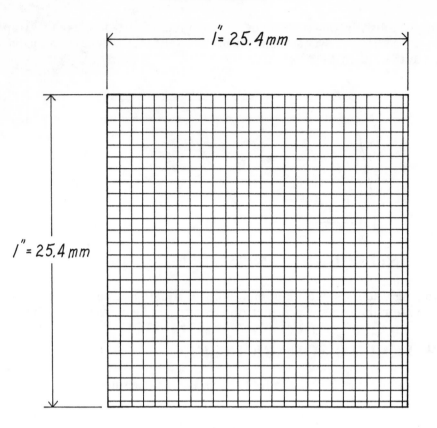

Therefore, to change square inches to square millimeters, multiply the number of square inches by 645.16.

1. 1 sq. in.   = 1 × 645.16 = 645.16 mm²
2. 1000 sq. in. = 1000 × 645.16 = 645 160 mm²
3. 2304 sq. in. = 2304 × 645.16 = 1 486 499 mm² (rounded)

SQUARE MILLIMETERS TO SQUARE INCHES

To convert square millimeters to square inches, divide the number of square millimeters by 645.16.

1. 645.16 mm²   = 645.16 ÷ 645.16 = 1 sq. in.
2. 6 000 mm²   = 6000 ÷ 645.16 = 9.3 sq. in. (rounded)
3. 500 000 mm² = 500000 ÷ 645.16 = 775 sq. in. (rounded)

FROM	TO	DO
Square Feet	Square Inches	× 144
Square Inches	Square Feet	÷ 144
Square Inches	Square Millimeters	× 645.16
Square Millimeters	Square Inches	÷ 645.16

KEY TERMS FOR WELDERS
PERIMETER
AREA
SQUARE
SQUARE UNITS
RECTANGLE
PARALLELOGRAM
TRAPEZOID

Unit 17—Practicing Four-Sided Measurement

Show all your work. Be certain the columns line up. Box your answers.

1. Convert the following square feet to square inches (to the nearest square inch).

   (a) 37 sq. ft.

   (b) 115 1/2 sq. ft.

   (c) 9.72 sq. ft.

   (d) 3.694 sq. ft.

   (e) 0.54 sq. ft.

   (f) 1,000 sq. ft.

2. Convert the following square inches to square feet (to the nearest tenth).

   (a) 81 sq. in.

   (b) 256 3/4 sq. in.

   (c) 15,984 sq. in.

   (d) 117.36 sq. in.

   (e) 100,000 sq. in.

   (f) 52 sq. in.

3. Convert the following square inches to square millimeters (to the nearest millimeter).

   (a) 90 sq. in.

   (b) 654.8 sq. in.

   (c) 137 1/8 sq. in.

   (d) 1,008 sq. in.

   (e) 36 sq. in.

   (f) 15.5 sq. in.

4. Convert the following square millimeters to square inches (to the nearest tenth).

   (a)   1 550 mm²   (b)   5 161.28 mm²

   (c)   100 mm²   (d)   9 675.5 mm²

   (e)   22 941 mm²   (f)   500 000 mm²

5. A customer orders 32 pieces of the following plate. What is the total area of the plates?

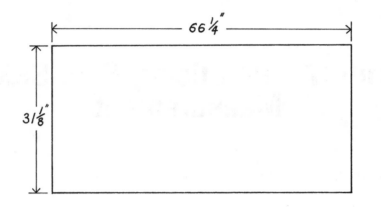

6. Find the permimeter and area of the following L-shaped plate.

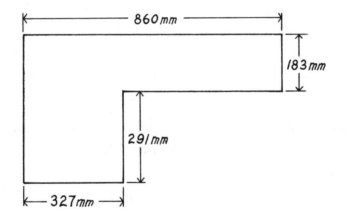

7. Find the area of the shaded part of the figure below. Express your answer in square feet and square inches.

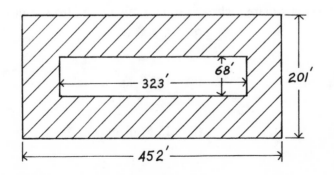

8. Find the perimeter and area of the following parallelogram.

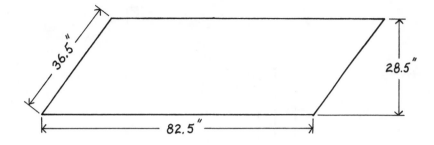

9. Find the perimeter and area of the following trapezoid.

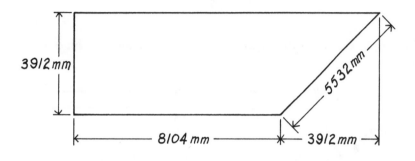

10. Find the area of the following trapezoid.

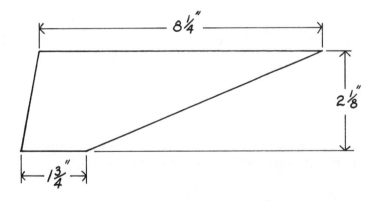

11. Find the area of scrap remaining after the trapezoid is cut from the parallelogram.

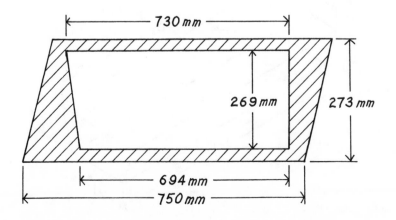

12. A heat exchanger has 16 panels of the following size. Find the total surface area of the panels. (Think about this problem.)

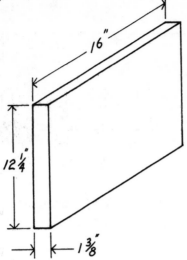

13. The following plate consists of a trapezoid and a parallelogram. Find the perimeter and area of the plate.

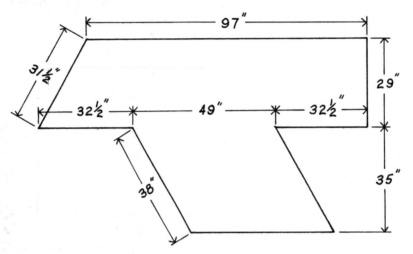

14. All surfaces of this sheet metal trough are to be painted with a rust-inhibiting paint. What is the total surface area, including the bottom, to be painted?

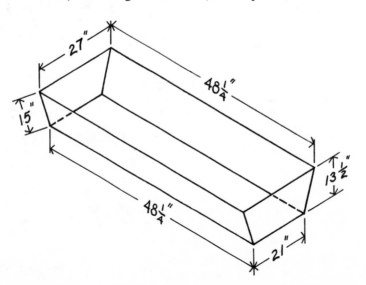

TRIANGULAR MEASURE

INTRODUCTION

The shape explained in this unit is the three-sided figure called the **triangle.** Triangles can be divided into a number of types, but this study will be limited to the three types of triangles that will appear most frequently in your welding work.

RIGHT TRIANGLE

A **right triangle** is a triangle in which one angle is 90°.

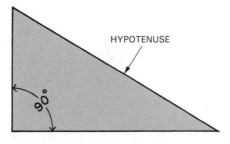

The sloping line is called the **hypotenuse** and the other two sides are referred to as sides. Laying out a right angle is a job you may be called upon to do in the shop or on the job site. One method is to form a **3-4-5 triangle.** A triangle with hypotenuse and sides of these lengths will produce a right angle triangle. Any multiples of these numbers will work, such as 6-8-10, or 9-12-15, or 30-40-50, etc. The longest length will, of course, be the hypotenuse.

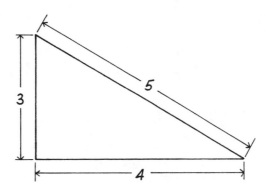

Another method is to select two pieces of the same length for the sides, say 10′. To calculate the hypotenuse, multiply the length by 1.414. In this example, the hypotenuse would be 14.14′. This hypotenuse, when assembled with the two sides, will produce a right angle triangle.

EQUILATERAL TRIANGLE

An **equilateral triangle** is a triangle in which all sides are the same length and all angles are equal to 60°. An equilateral triangle is one of the strongest shapes in nature and is often used in fabricating.

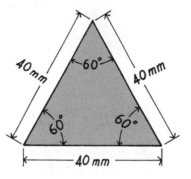

ISOSCELES TRIANGLE

An **isosceles triangle** is a triangle in which two of the three sides are of equal length and two of the three angles are equal.

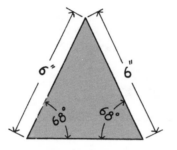

PERIMETER OF A TRIANGLE

To calculate the perimeter of each of these three types of triangles, simply add the three lengths or sides.

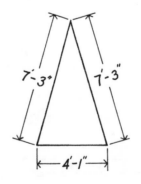

P = 7 ft. 3 in. + 7 ft. 3 in. + 4 ft. 1 in.
P = 18 ft. 7 in.

AREA OF A TRIANGLE

To calculate the area of each of the three types of triangles, multiply 1/2 times the base line times the height of the triangle.

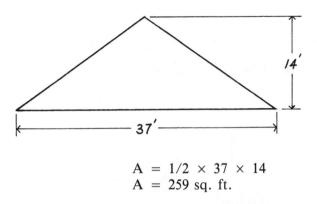

A = 1/2 × 37 × 14
A = 259 sq. ft.

It is interesting to note that the total of the three angles within any triangle is equal to 180°. So, if you know two of the angles, you can easily calculate the third.

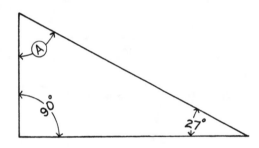

Angle A = 180° − 117°
A = 63 °

KEY TERMS FOR WELDERS
TRIANGLE
RIGHT TRIANGLE
3-4-5 TRIANGLE
HYPOTENUSE
EQUILATERAL TRIANGLE
ISOSCLES TRIANGLE

Unit 18—Practicing Triangular Measurement

Show all your work. Box your answers.

1. Find the perimeter and area of this right triangle.

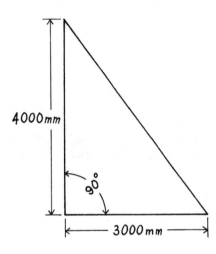

2. Find the perimeter and area of this isosceles triangle.

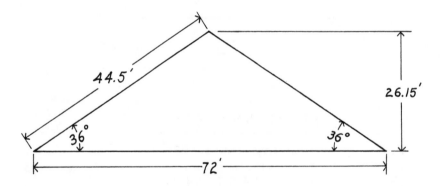

3. Find the perimeter and area of this equilateral triangle.

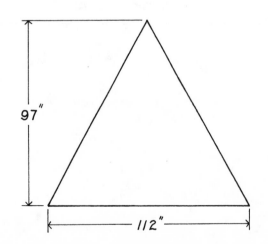

4. A rectangular hole is cut from this equilateral triangle. What is the remaining area of the triangle?

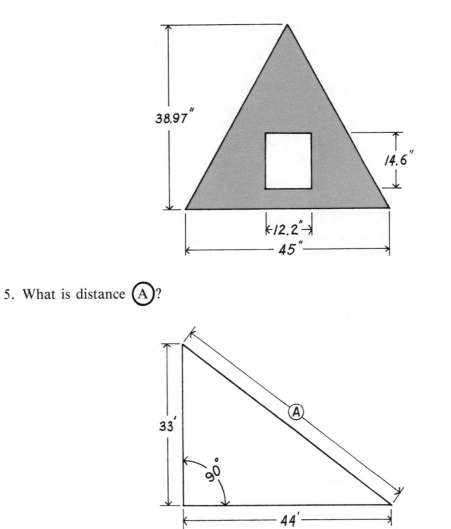

5. What is distance Ⓐ?

6. Find the total area of these nine equilateral triangles.

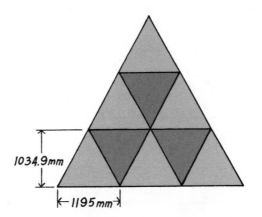

7. Calculate the perimeter and area of the following triangle.

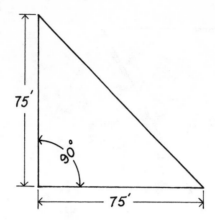

8. The frame for the roof shown below is made by welding 3 in. S-beam.
   (a) What is the total length of beam used?
   (b) What is the total area of the roof?

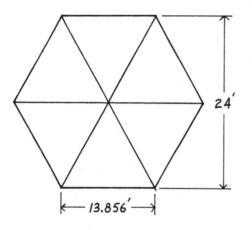

9. Calculate the following for this structural frame.
   (a) Total length of framing required.
   (b) Total area of the shape.

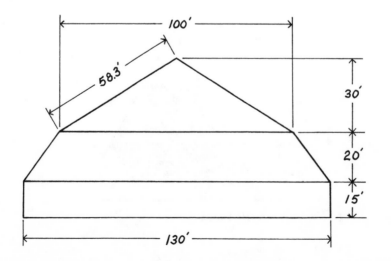

10. What is the total length of angle iron used in the following roof truss?

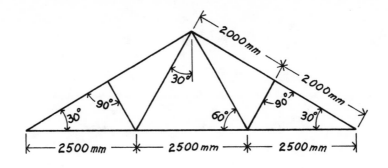

CIRCULAR MEASURE

INTRODUCTION

Working with circles and circular features will present no problem to the welder familiar with the basic mathematic characteristics of the circle. To begin, here is the terminology you will need to know:

1. The **circumference** is the distance around a circle.
2. The **diameter** is the straight line distance across a circle and passing through the center.
3. The **radius** is the straight line distance from the center to the edge of a circle.

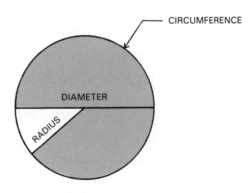

If you know certain characteristics of a circle, you can calculate other characteristics.

CIRCUMFERENCE OF A CIRCLE

If you know the diameter of a circle, the circumference can be calculated from the following formula:

$$C = \pi \times D$$
$$\text{Circumference} = \pi \times \text{Diameter}$$

This formula requires some explanation. The symbol π is actually a letter from the Greek language and it represents the number 3.14. It is spelled "pi," but is pronounced "pie." By multiplying the diameter of a circle by this number, you will arrive at the circumference.

So, the diameter and circumference are related to each other through the number π.

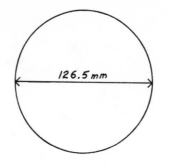

$$\begin{aligned} \text{Circumference} &= \pi \times \text{D} \\ \text{C} &= 3.14 \times 126.5 \\ \text{C} &= 397.21 \text{ mm} \end{aligned}$$

DIAMETER OF A CIRCLE

If the circumference is known, the diameter of a circle can be calculated from the following formula:

$$\text{D} = \frac{\text{C}}{\pi}$$

$$\text{Diameter} = \frac{\text{Circumference}}{\pi}$$

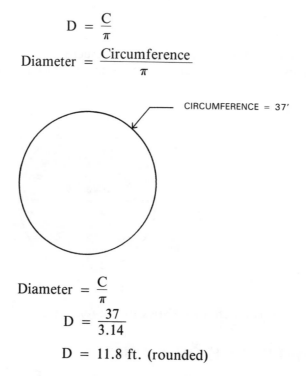

CIRCUMFERENCE = 37'

$$\text{Diameter} = \frac{\text{C}}{\pi}$$

$$\text{D} = \frac{37}{3.14}$$

$$\text{D} = 11.8 \text{ ft. (rounded)}$$

AREA OF A CIRCLE

The area of a circle can be calculated if the radius is known.

$$\begin{aligned} \text{A} &= \pi \times \text{R} \times \text{R} \\ \text{Area} &= \pi \times \text{Radius} \times \text{Radius} \end{aligned}$$

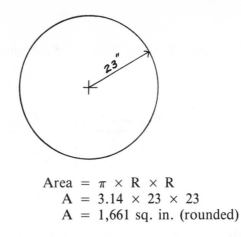

Area = $\pi \times R \times R$
A = 3.14 $\times$ 23 $\times$ 23
A = 1,661 sq. in. (rounded)

The formulas for circumference, diameter, and area are extremely important and you should memorize them.

CIRCULAR SHAPES

Many fabricated objects consist of partial circles and straight lines. By recognizing basic shapes within these objects, some important characteristics can be calculated. These three examples shown here will give you some guidance.

HALF CIRCLE

Given the information on the drawing, the perimeter and area can be calculated.

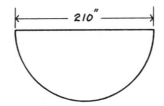

PERIMETER OF A HALF CIRCLE

The perimeter is equal to the straight line length of 210 in. plus one half the circumference of a circle with a diameter of 210 in.

$$C = \pi \times D$$

Circumference (half circle) $= \dfrac{\pi \times D}{2} = \dfrac{3.14 \times 210 \text{ in.}}{2} = 329.7$ in.

Perimeter (half circle) = 329.7 + 210 = 539.7 in.

AREA OF A HALF CIRCLE

The area is one half the area of a circle with a radius of 105 in.

Area (half circle) $= \dfrac{\pi \times R \times R}{2} = \dfrac{3.14 \times 105 \text{ in.} \times 105 \text{ in.}}{2} = 17,309.25$ sq. in.

CYLINDER

Given the information on the drawing, the area of the curved surface and the area of the entire surface can be calculated.

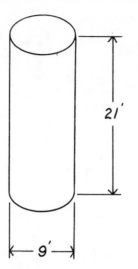

AREA OF CURVED SURFACE

Calculate the circumference and redraw the curved surface as a flat rectangle; then, calculate the area.

$$C = \pi \times D$$
$$C = 3.14 \times 9' = 28.26 \text{ ft.}$$

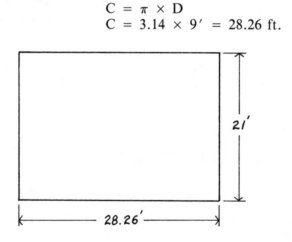

$$\text{Area} = 21' \times 28.26' = 593.46 \text{ sq. ft.}$$

AREA OF ENTIRE SURFACE OF A CYLINDER

Add the area of the two circular ends to the area of the curved surface.

$$
\begin{aligned}
\text{Area (of both ends)} \quad &= \pi \times R \times R \times 2 \\
&= 3.14 \times 4.5 \times 4.5 \times 2 \\
&= 127.17 \text{ sq. ft.}
\end{aligned}
$$

$$
\begin{aligned}
\text{Area of entire surface} \quad &= 127.17 + 593.46 \\
&= 720.63 \text{ sq. ft.}
\end{aligned}
$$

SEMI-CIRCULAR SIDED SHAPE

Given the information on the drawing, the perimeter and area can be calculated.

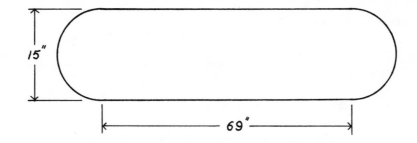

PERIMETER OF A SEMI-CIRCULAR SIDED SHAPE

Calculate the perimeter by adding the two straight sides and the semi-circular ends. Since the two ends together equal one circle, the calculations are as follows:

$$C = \pi \times D$$
$$C = 3.14 \times 15' = 47.1 \text{ in.}$$
$$\text{Perimeter} = 69 + 69 + 47.1 = 185.1 \text{ in.}$$

AREA OF A SEMI-CIRCULAR SIDED SHAPE

The area consists of two semicircles (one complete circle), plus a rectangle.

Area of circle	$= \pi \times R \times R = 3.14 \times 7.5 \times 7.5$
	$= 176.625 \text{ sq. in.}$
Area of rectangle	$= 69 \times 15 = 1,035 \text{ sq. in.}$
Area of shape	$= 176.625 + 1,035 = 1,211.625 \text{ sq. in.}$

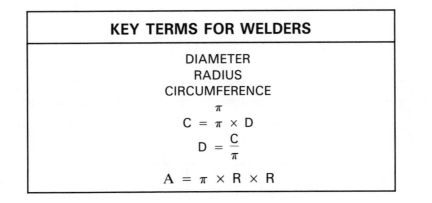

Unit 19—Practicing Circular Measurement

Show all your work. Box your answers.

1. Calculate the area of scrap remaining after the two circles are cut from this piece of sheet metal to the nearest inch.

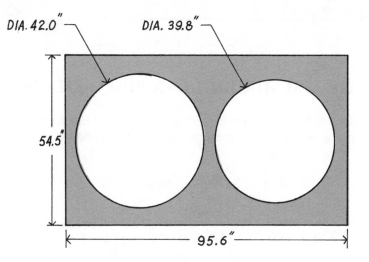

2. Find the area to the nearest square foot and the perimeter to the nearest inch of the shape shown below.

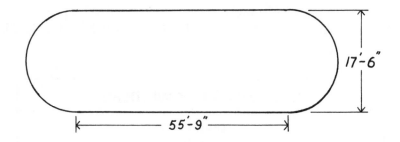

3. Calculate the length of strapping needed for this pipe hanger to the nearest sixteenth of an inch.

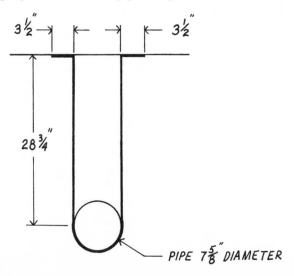

PIPE $7\frac{5}{8}''$ DIAMETER

4. Calculate the area and perimeter of this flame-cut piece to the nearest sixteenth of an inch.

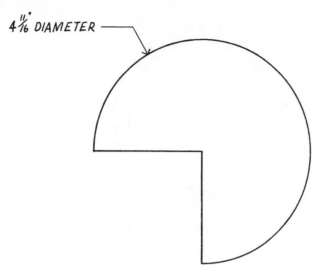

$4\frac{11}{16}"$ DIAMETER

5. Find the area of the plate before and after the holes have been cut to the nearest inch.

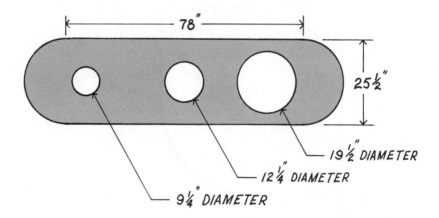

78"

$25\frac{1}{2}"$

$19\frac{1}{2}"$ DIAMETER

$12\frac{1}{4}"$ DIAMETER

$9\frac{1}{4}"$ DIAMETER

6. Calculate the total area of this cylinder to the nearest inch. Convert your answer to the nearest millimeter.

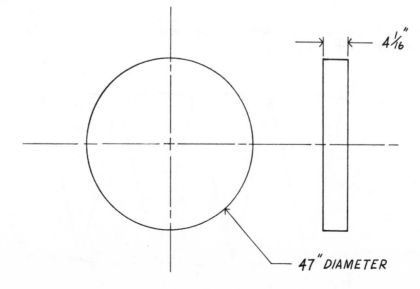

$4\frac{1}{16}"$

47" DIAMETER

7. Find the area of this steel ring to the nearest mm. Convert your answer to the nearest tenth of a square inch.

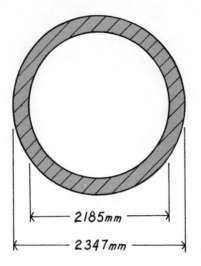

2185mm

2347mm

8. Find the total length to the nearest tenth of an inch of strapping needed for 15 pipe corner braces.

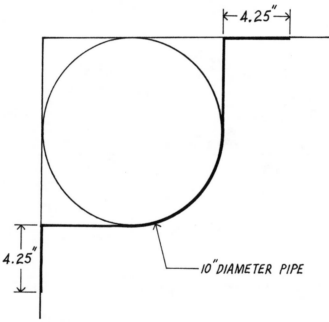

4.25"

4.25"

10" DIAMETER PIPE

9. Find the total length of curved pieces and the total length of straight pieces needed for 20 of these wrought iron sections.

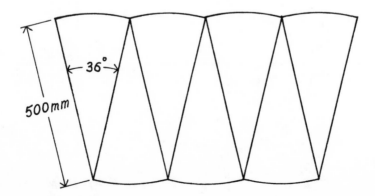

36°

500mm

10. Find the area and perimeter of this shape to the nearest inch.

Section Five

VOLUME, WEIGHT, AND BENDING METAL

Objectives for Section Five, Volume, Weight, and Bending Metal

After studying this section, you will be able to:

- Define regular shaped objects
- Calculate the volume of regular shaped objects
- Convert volume measurements between units
- Calculate the volume of irregular shaped objects
- Compute weight calculations based on volume
- Compute weight calculations based on length
- Define four types of metal bends
- Define inside and outside corner
- Compute stock length for various metal bends

Contents for Section Five, Volume, Weight, and Bending Metal

VOLUME MEASURE

INTRODUCTION

Volume is the amount of space an object occupies. All objects exist in three dimensions: length, width, and height. The volume of an object is the product of these dimensions and is expressed in cubic units. For example:

> 82 cubic inches
> 119 cubic feet
> 1 789 mm³
> 37.5 cu. ft.
> 10 cu/in.

VOLUME OF REGULAR SHAPED OBJECTS

The volume of an object having a uniform cross-section is determined by calculating the area of the cross-section and then multiplying by the length.

$$V = A \times L$$

CUBE

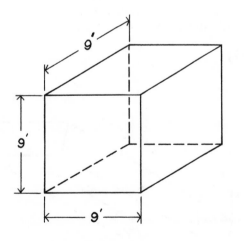

Area of cross-section: 9 × 9 = 81
Volume: 81 × 9 = 729 cu. ft.

SOLID RECTANGLE

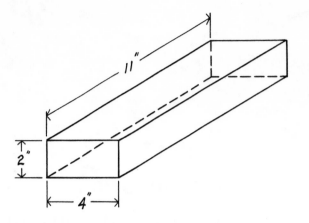

Area of cross-section: $2 \times 4 = 8$
Volume: $8 \times 11 = 88$ cu. in.

SOLID PARALLELOGRAM

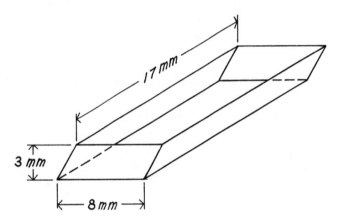

Area of cross-section: $3 \times 8 = 24$
Volume: $24 \times 17 = 408$ mm³

SOLID TRAPEZOID

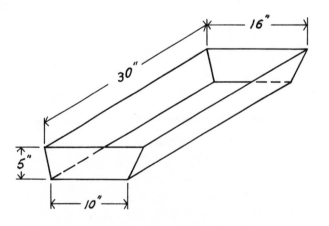

Area of cross-section: $16 + 10 = 26$
$26 \times 5 = 130$
$130 \div 2 = 65$
Volume: $65 \times 30 = 1,950$ cu. in.

SOLID TRIANGLE

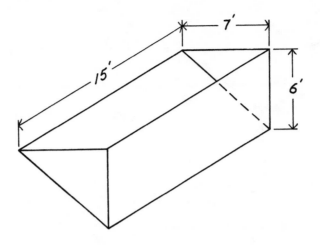

Area of cross-section: 7 × 6 × 1/2 = 21
Volume: 21 × 15 = 315 cu. ft.

CYLINDER

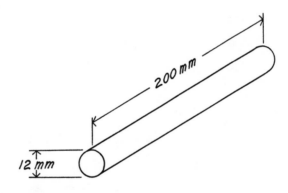

Area of cross-section: 3.14 × 6 × 6 = 113.04
Volume: 113.04 × 200 = 22 608 mm³

SOLID HALF CIRCLE

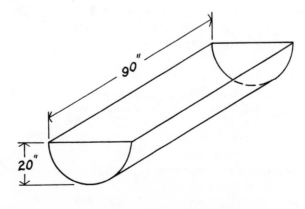

Area of cross-section: $\dfrac{3.14 \times 20 \times 20}{2} = 628$
Volume: 628 × 90 = 56,520 cu. in.

SOLID SEMI-CIRCULAR SIDED SHAPE

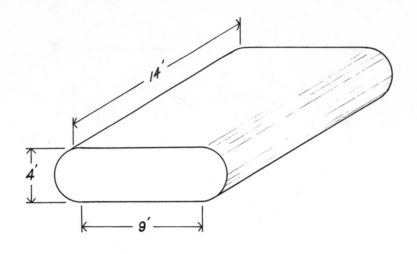

Area of cross-section: $3.14 \times 2 \times 2 = 12.56$
$+ 9 \times 4 \qquad = \underline{36.}$
$\qquad\qquad\qquad\qquad\qquad 48.56$

Volume: $14 \times 48.56 = 679.84$ cu. ft.

CONVERTING VOLUME MEASUREMENTS

Linear dimensions of objects are usually expressed in feet, inches, or millimeters. Volume, as illustrated in this unit, can be expressed in cubic feet, cubic inches, or cubic millimeters. However, it is quite often more useful to express the volume (or capacity) of an object in gallons or liters.

A **liter** is a metric measurement of capacity. It is a little bigger than a quart. It is important for you to be able to convert back and forth between the various units of volume. Listed below are some common equivalent volume measurements:

1 gallon (USA) = 231 cu. in.
1 cubic foot   = 7.48 gallons (USA)
1 gallon (USA) = 3.785 liters
1 cubic foot   = 1,728 cu. in.
1 cubic inch   = 16 387.064 mm³

The degree of accuracy usually required in a welding shop allows the use of rounded figures.

1 cubic inch   = 16 387 mm³
1 gallon (USA) = 3.79 liters

Also, since conversions often result in lengthy decimals, it is normal practice to round the answer.

1. GALLONS TO CUBIC INCHES
   Multiply the number of gallons by 231.
   40 gallons = 40 × 231 = 9,240 cu. in.

2. CUBIC INCHES TO GALLONS
   Divide the number of cubic inches by 231.
   2,425.5 cu. in. = 2,425.5 ÷ 231 = 10.5 gallons.

3. GALLONS TO CUBIC FEET
   Divide the number of gallons by 7.48.
   1,000 gallons = 1,000 ÷ 7.48 = 133.6898 cu. ft.
                                       = 133.69 cu. ft. (rounded).

4. CUBIC FEET TO GALLONS
   Multiply the number of cubic feet by 7.48.
   16 cu. ft. = 16 × 7.48 = 119.68 gallons.

5. GALLONS TO LITERS
   Multiply the number of gallons by 3.79.
   5 gallons = 5 × 3.79 = 18.95 liters.

6. LITERS TO GALLONS
   Divide the number of liters by 3.79.
   85 liters = 85 ÷ 3.79 = 22.4274 gallons.
                           = 22.43 gallons (rounded).

7. CUBIC MILLIMETERS TO CUBIC INCHES
   Divide the number of cubic millimeters by 16 387.
   27 000 000 mm³ = 27 000 000 ÷ 16,387 = 1,647.64 cubic inches (rounded).

8. CUBIC INCHES TO CUBIC MILLIMETERS
   Multiply the number of cubic inches by 16,387.
   9 cubic inches = 9 × 16,387 = 147 483 mm³.

9. CUBIC FEET TO CUBIC INCHES
   Multiply the number of cubic feet by 1,728.
   27 cu. ft. = 27 × 1,728 = 46,656 cu. in.

10. CUBIC INCHES TO CUBIC FEET
    Divide the number of cubic inches by 1,728.
    5,000 cubic inches = 5,000 ÷ 1,728 = 2.8935 cu. ft.
                                          = 2.89 cu. ft. (rounded).

VOLUME OF IRREGULAR SHAPED OBJECTS

The volume of many irregular shaped objects can be easily calculated. The way to compute this is to first divide the object into regular shaped parts, then, calculate the volume of each part; and finally, add all the volumes. The example below calculates the volume to the nearest tenth of an inch. The material is two in. thick and all holes are two in. in diameter.

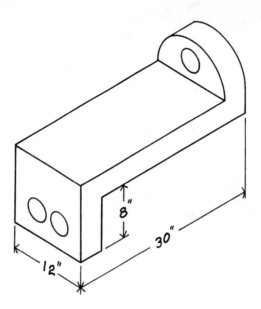

Step One: Divide the object into regular shaped parts.

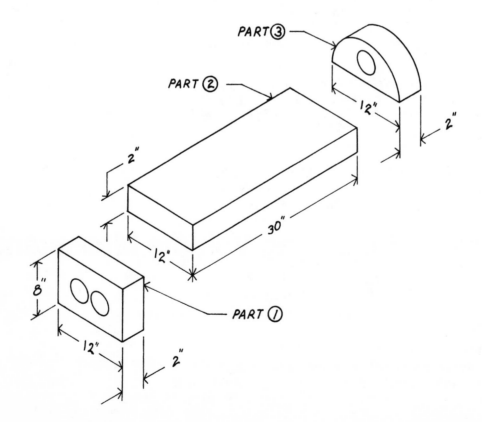

Step Two: Calculate the volume of each part.

Part 1

Gross volume	$= 2 \times 8 \times 12$	$=$	192.00
Minus volume of holes	$= \pi \times 1 \times 1 \times 2 \times 2$	$=$	$-$ 12.56
Volume	$=$		179.44 cu. in.

Part 2

Volume $= 2 \times 12 \times 30 =$ 720.00 cu. in.

Part 3

Volume of half circle $= \dfrac{\pi \times 6 \times 6 \times 2}{2}$ $=$ 113.04

Minus volume of hole $= \pi \times 1 \times 1 \times 2$ $=$ $-$ 6.28

Volume $=$ 106.76

Step Three: Add the volume together.

Part 1	179.44
Part 2	720.00
Part 3	106.76
Total Volume	1,006.2 cu. in.

A WORD OF CAUTION

One of the difficulties encountered in calculating volumes, especially of irregular shaped objects, is keeping "track" of your figures. As pointed out in Chapter one, it is crucial that you form the habit of being organized in your math work. You will find that math concepts do not become more difficult. However, the sheer volume of figures you will be dealing with may cause confusion. So, one of the main challenges you will continually face is to maintain order and control of your math work.

KEY TERMS FOR WELDERS
VOLUME
$V = A \times L$
LITER

Unit 20—Practicing Volume Measurement

Show all your work. Be certain the columns line up. Box your answers.

1. Calculate the following conversions:
   (a) How many cu. in. of space does 4 1/2 gallons of water occupy?

(b) How many cubic feet are contained in 29 000 000 cubic millimeters? Calculate to the nearest tenth.

(c) A swimming pool contains 1,800 cu. ft. of water. How many gallons does the pool contain? Calculate to the nearest gallon.

(d) Convert 1.125 cu. in. to mm³. Calculate to the nearest mm.

(e) A home heating oil storage tank holds 450 liters of oil. How many gallons does it hold? Calculate to the nearest gallon.

(f) A microwave oven has a volume of 2,475 cu. in. What is the volume in cubic feet? Calculate to the nearest tenth.

(g) How many liters are there in a 20 gallon gasoline tank? Calculate to the nearest tenth.

(h) How many cubic inches of space are there in a refrigerator with a capacity of 8 cubic feet?

(i) The water tank for the town of South Windsor has a capacity of 500,000 gallons. How many cubic feet of water does it contain? Calculate to the nearest cubic foot.

(j) A motorcycle engine has a size of 45 cu. in. What is the size of the engine in mm³?

(k) Convert 8,000 cu. in. to cu. ft. Calculate to the nearest tenth.

(l) A quenching tank measures 2′ × 3′ × 6′. How many gallons does it hold? Calculate to the nearest hundredth.

(m) How many gallons are there in 1,000 cubic inches? Calculate to the nearest thousandth.

(n) A block of precision steel contains 2 539 985 cubic millimeters. What is the volume in cubic inches?

(o) Convert 35,000 gallons to cubic feet. Calculate to the nearest tenth.

2. How many cubic millimeters are contained in the following piece of steel? Calculate to the nearest hundredth mm.

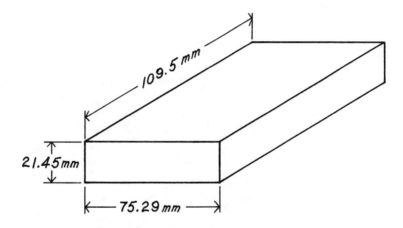

3. What is the volume of steel in this triangular rod? Calculate to the nearest cubic inch.

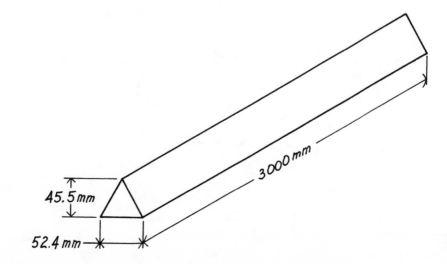

4. The concrete foundation for a newly installed press had the following dimensions. How many cubic yards of concrete were used (to the nearest one quarter of a yard)?

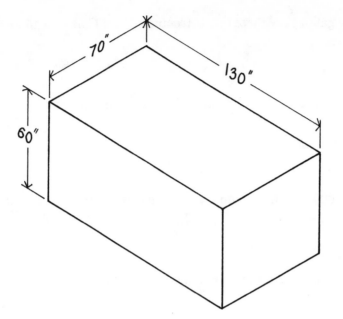

5. Find the total volume of this weldment to the nearest half cubic inch.

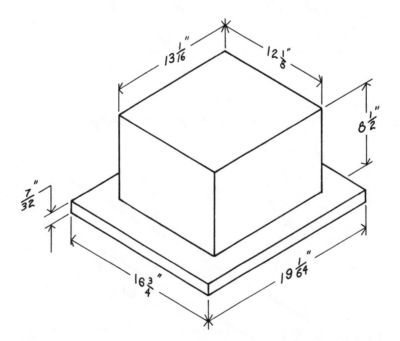

6. This block of steel is turned on a lathe to a diameter of 18 in. What volume of material is removed (to the nearest hundredth of a cubic inch)?

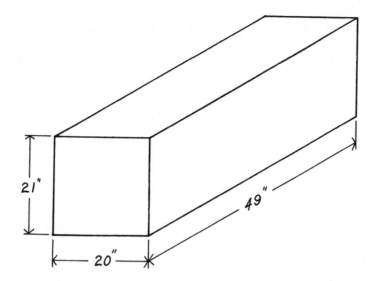

7. Find the volume of this shape.

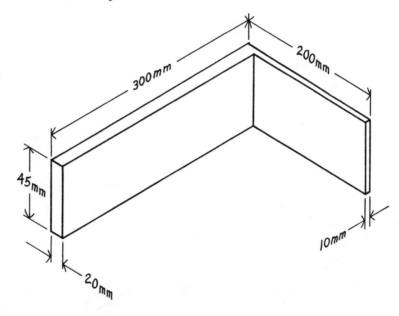

8. Twenty joints are welded as illustrated. Fillet welds on both sides of the joint are similar. What volume of weld material is deposited?

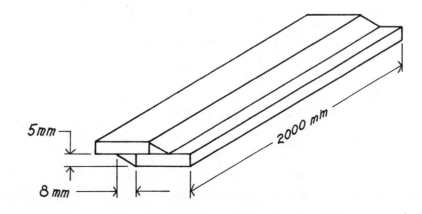

9. The ventilating system in this fabricating shop can evacuate 5,200 cu. ft. of air per minute. How long will it take to exchange all of the air in the shop? Calculate to the nearest minute.

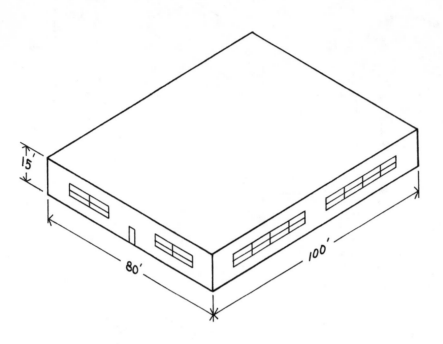

10. What is the total volume of cast iron in this weldment? Express your answer in cubic inches.

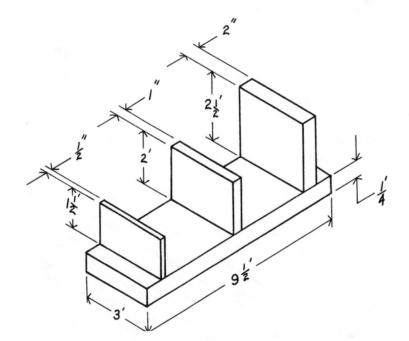

11. Calculate the volume of this silo to the nearest cubic foot.

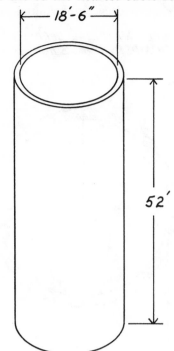

12. How many cu. ft. of steel are in this 250 ft. roll?

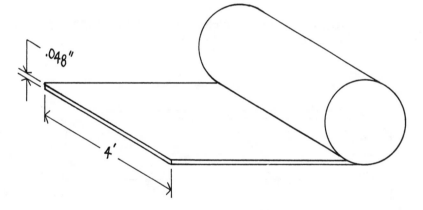

13. Seven bins are to be constructed as shown below. What is the total volume of the bins to the nearest cu. ft.?

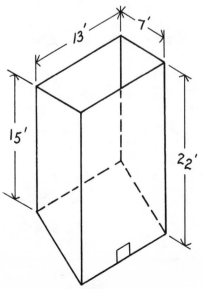

14. The worn surface of this 450 mm long shaft is built up by surfacing. Three millimeters of material are deposited all around. What total volume of weld material is deposited?

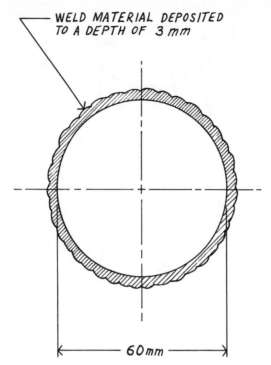

WELD MATERIAL DEPOSITED TO A DEPTH OF 3 mm

60mm

15. This 15 ft. wide above-ground pool is filled to within 6 in. of the top. How many liters of water does it contain (to the nearest ten liters)?

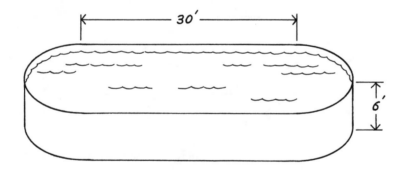

30′

6′

16. Which container has the greater capacity and by how much to the nearest gallon?

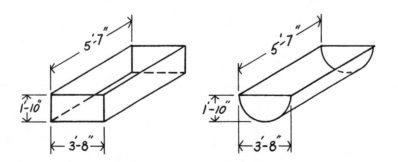

5′-7″   5′-7″

1′-10″   1′-10″

3′-8″   3′-8″

17. All material in this weldment is 1/4 in. thick unless otherwise shown on the drawing. The flame cut hole is 10 in. in diameter. The triangular support pieces are only on one side of the upright. Find the total volume to the nearest quarter of a cubic inch.

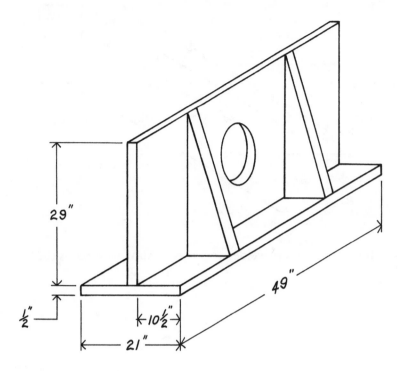

18. Find the volume of each of the pipes to the nearest cu. in.

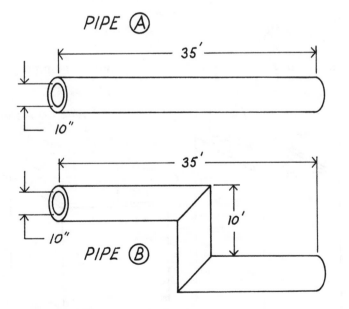

19. How many gallons will this trough hold? Calculate to the nearest gallon.

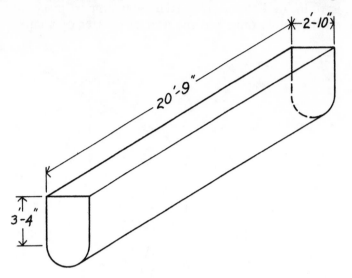

20. Two plates, 3 000 mm long, are welded together as shown. What is the total volume of weld deposited?

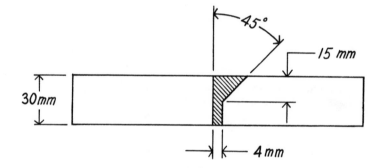

21. Calculate the volume of the two settling tanks, including the pipes, to the nearest gallon.

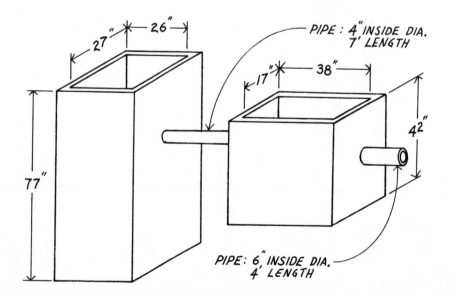

WEIGHT MEASURE

INTRODUCTION

When calculating the weight of weldments, you will encounter one of the following conditions. The weldment may consist of randomly shaped metal plates or standard structural shapes (such as angle iron, channel, etc.) or a combination of both. The calculations for randomly shaped plates are based on volume. The calculations for standard structural shapes are based on length.

WEIGHT CALCULATION BASED ON VOLUME

The weight of a block of steel is easy to determine if you know the volume of the piece and the weight of the steel per unit volume. An example will clarify this. Steel has a weight of 0.283 lbs. per cu. in.; therefore, a block of steel with a volume of 250 cu. in. weighs 250 × 0.283 or 70.75 lbs.

Metals have different densities and therefore different weights as the following list indicates. Even the weight of different classifications of steel will vary depending on the composition of the steel.

WEIGHT OF METALS		
	lb/in³	g/cm³
Magnesium	0.0628	1.738
Aluminum	0.0975	2.699
Zinc	0.2570	7.114
Tin	0.2633	7.288
Cast Iron	0.2665	7.377
Steel	0.2835	7.847
Copper	0.3210	8.885
Lead	0.4096	11.338
Tungsten	0.6900	19.099
Gold	0.6969	19.290

You will notice the above list introduces a new unit of measure, namely, g/cm³. This means grams per cubic centimeter. Grams (g) and centimeters (cm) are metric measurements commonly used in these calculations.

The **gram** (g) is a measure of weight and is a very small measure indeed. A paper clip, for example, weighs about one gram. Since many objects handled in everyday life are much heavier than a few grams, it is common practice to use 1 000 grams as a unit of measure. One thousand grams is called a **kilogram** and is a little more than two pounds. You will see it written in two other styles, either as **kg** or as **kilo.** To convert from grams to kilograms, divide by 1 000. To convert from kilograms to grams, multiply by 1 000.

The **centimeter** (cm) is a measure of length and is ten millimeters long. A cube of sugar, for example, measures about 1 cm × 1 cm × 1 cm and is therefore about 1 cm³ in volume.

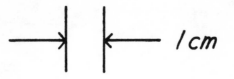

Because the millimeter is so small, many objects are measured in centimeters. To convert from millimeters to centimeters, divide by 10. To convert from centimeters to millimeters, multiply by 10.

WEIGHT CALCULATION BASED ON LENGTH

A full description of the standard structural shapes produced by American steel makers is provided in the *Manual of Steel Construction* issued by the American Institute of Steel Construction Inc. The Canadian equivalent is produced by the Canadian Institute of Steel Construction. Among the statistics provided for each shape is the weight per unit length. A summary of selected shapes is listed below.

You will notice that the figures may be expressed in the Standard system as pounds per foot or in the Metric system as kilograms per meter. As previously explained, one kilogram equals 1 000 grams. Similarly, one **meter** (m) is equal to 1 000 millimeters and is a bit longer than one yard. You will see it written as **m.** To convert from millimeters to meters, divide by 1 000. To convert from m to mm, multiply by 1 000.

The weight of a standard structural shape is easy to determine if you know the length of the piece and the weight of the shape per unit length. For example, if a 4 in. channel weighs 5.4 lbs. per foot, the weight of a 6′ piece would be 5.4 × 6 or 32.4 lbs.

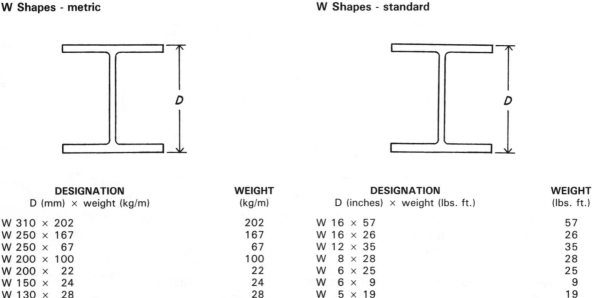

W Shapes - metric

DESIGNATION D (mm) × weight (kg/m)	WEIGHT (kg/m)
W 310 × 202	202
W 250 × 167	167
W 250 × 67	67
W 200 × 100	100
W 200 × 22	22
W 150 × 24	24
W 130 × 28	28

W Shapes - standard

DESIGNATION D (inches) × weight (lbs. ft.)	WEIGHT (lbs. ft.)
W 16 × 57	57
W 16 × 26	26
W 12 × 35	35
W 8 × 28	28
W 6 × 25	25
W 6 × 9	9
W 5 × 19	19

Weight Measure

S Shapes - metric

DESIGNATION D (mm) × weight (kg/m)	WEIGHT (kg/m)
S 610 × 158	158
S 250 × 52	52
S 200 × 34	34
S 180 × 30	30
S 150 × 26	26
S 130 × 22	22
S 75 × 11	11

S Shapes - standard

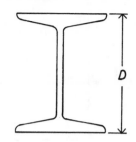

DESIGNATION D (inches) × weight (lbs. ft.)	WEIGHT (lbs. ft.)
S 18 × 54.7	54.7
S 12 × 50	50.0
S 12 × 40.8	40.8
S 8 × 23	23.0
S 8 × 18.4	18.4
S 6 × 17.25	17.25
S 3 × 5.7	5.7

Channels (C Shapes) - metric

DESCRIPTION D (mm) × weight (kg/m)	WEIGHT (kg/m)
C 250 × 37	37
C 200 × 28	28
C 180 × 18	18
C 150 × 19	19
C 130 × 17	17
C 130 × 13	13
C 100 × 9	9

Channels (C Shapes) - standard

DESCRIPTION D (inches) × weight (lbs. ft.)	WEIGHT (lbs. ft.)
C 10 × 30	30.0
C 9 × 13.4	13.4
C 8 × 18.75	18.75
C 8 × 11.5	11.5
C 6 × 10.5	10.5
C 5 × 9	9.0
C 4 × 5.4	5.4

ANGLES (L Shapes) - metric

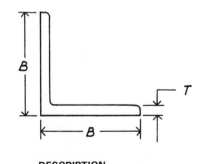

DESCRIPTION B (mm) × B (mm) × T (mm)	WEIGHT (kg/m)
L 200 × 200 × 16	48.216
L 150 × 100 × 13	24.257
L 125 × 90 × 13	20.685
L 100 × 100 × 16	23.685
L 90 × 75 × 10	12.173
L 55 × 55 × 8	6.399
L 45 × 55 × 6	3.244

ANGLES (L Shapes) - standard

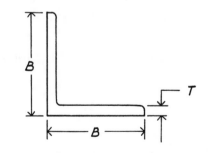

DESCRIPTION B (inches) × B (inches) × T (inches)	WEIGHT (lbs. ft.)
L 6 × 4 × 7/8	27.2
L 4 × 3 1/2 × 1/2	11.9
L 3 × 3 × 3/8	7.2
L 3 1/2 × 2 1/2 × 7/16	8.3
L 3 × 3 × 1/2	9.4
L 2 1/2 × 2 × 3/8	5.3
L 2 × 2 × 1/8	1.65

KEY TERMS FOR WELDERS
gram = g
kilogram = kg
centimeter = cm
meter = m
W Shapes
S Shapes
C Shapes
L Shapes

Unit 21—Practicing Weight Measurement

Show all your work. Be certain the columns line up. Box your answers.

1. Calculate the weight of this steel bar to the nearest kilogram. (HINT: Since the table of structural steel shapes provides the weight in g/cm³, it is best to calculate the volume in cubic centimeters. First, convert the millimeter dimensions to centimeters and then proceed with finding the volume in cubic centimeters.)

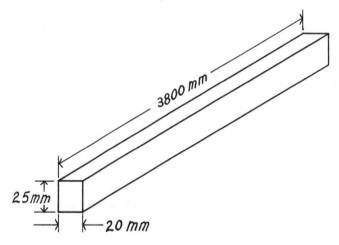

3800 mm

25mm

20 mm

2. Calculate the weight of the following copper plate to the nearest pound.

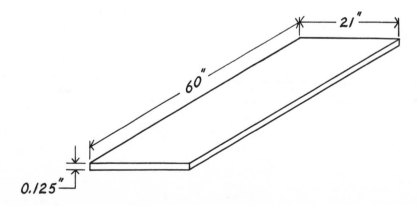

21"

60"

0.125"

3. Calculate the weight of the following cast iron bar to the nearest pound.

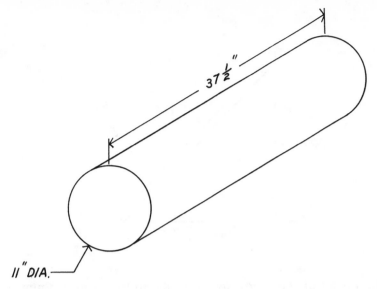

4. Calculate the weight of the following piece of angle iron to the nearest tenth of a kilogram. (NOTE: For standard structural shapes refer to the description provided in the text.)

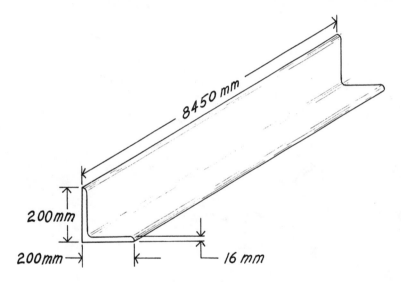

5. Calculate the weight of the following S shape to the nearest tenth of a kilo.

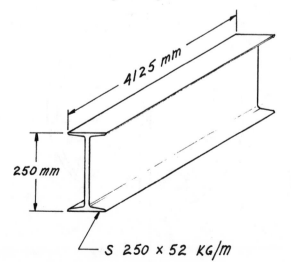

6. Calculate the weight of the following triangular piece of steel to the nearest tenth of a kilogram.

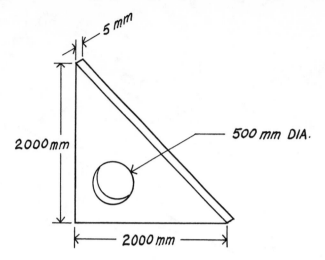

2000 mm

5 mm

500 mm DIA.

2000 mm

7. Calculate the weight of the following steel tube to the nearest pound.

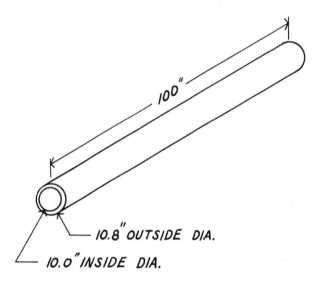

100"

10.8" OUTSIDE DIA.

10.0" INSIDE DIA.

8. Calculate the weight of this piece of steel to the nearest tenth of a pound. The diameter is 36 in. and the thickness is two in.

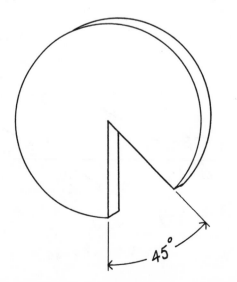

45°

9. If the price of gold is $380.00 per ounce, what is the value of the following bar of gold to the nearest cent?

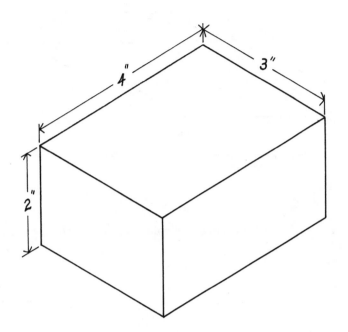

10. Calculate the weight of this steel hub and plate to the nearest tenth of a pound.

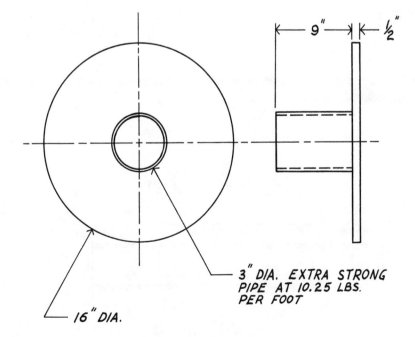

3" DIA. EXTRA STRONG PIPE AT 10.25 LBS. PER FOOT

16" DIA.

11. The following frame is made of square structural tubing weighing 47.9 lbs. per ft. What is the weight of the frame to the nearest tenth of a pound?

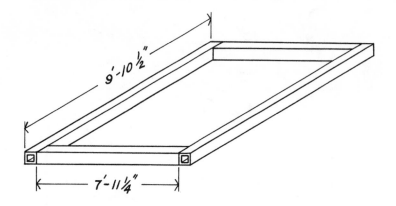

12. Calculate the weight of the following steel platform to the nearest kilogram.

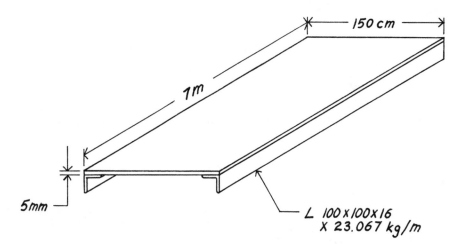

13. A customer order required sixteen columns as illustrated. What is the total weight of the order to the nearest pound?

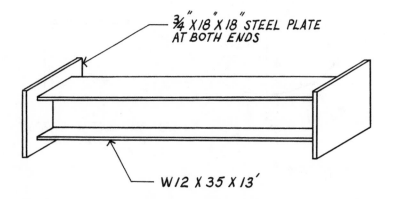

14. Calculate the weight of the following weldment to the nearest kilogram.

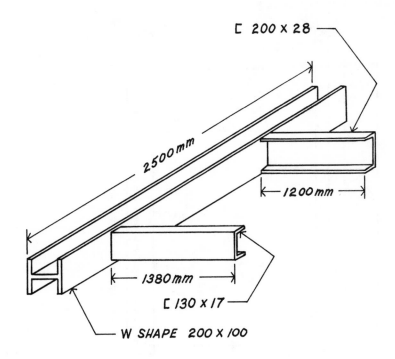

15. A machine base was fabricated from the following list of materials. What is the weight of the machine base to the nearest pound?

ITEM	QUANTITY	DESCRIPTION	SIZE	WT. PER FT.
g	11	SOLID ROUND BAR	1⅛" DIA. × 7"	3.38
f	1	S SHAPE	3" × 5.7 LBS. × 2'-6"	5.7
e	4	ANGLE IRON	3" × 3" × ⅜" × 17⅝"	7.2
d	6	STEEL PLATE	½" × 3" × 3"	–
c	4	ANGLE IRON	2½" × 2" × ⅜" × 13½"	5.3
b	2	ANGLE IRON	2" × 2" × ⅛" × 39"	1.65
a	1	STEEL PLATE	⅜" × 25" × 40"	–

16. Calculate the weight of the following pillar to the nearest pound.

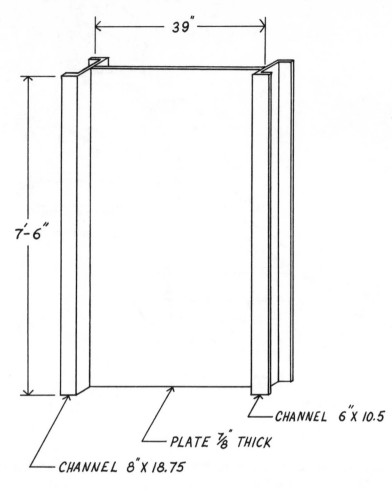

39"

7'-6"

CHANNEL 6"X 10.5

PLATE ⅞" THICK

CHANNEL 8"X 18.75

17. What is the weight of the following steel base for a swivel chair? Calculate your answer to the nearest gram.

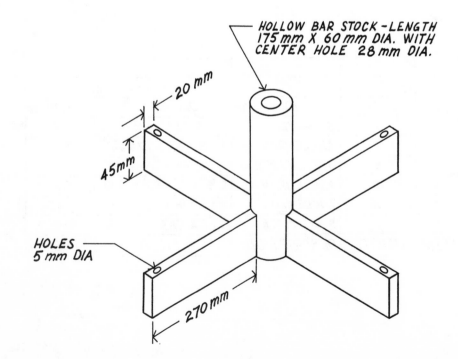

HOLLOW BAR STOCK – LENGTH 175 mm X 60 mm DIA. WITH CENTER HOLE 28 mm DIA.

20 mm

45mm

HOLES 5 mm DIA

270 mm

18. Calculate the weight of this section of steel fence to the nearest pound.

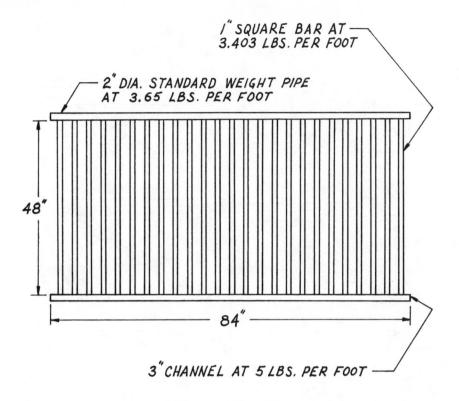

19. Calculate the weight of the following test piece to the nearest tenth of a pound. All plates are 3/8 in. thick and the overall length is 38 in. (HINT: A difficulty with this type of problem is keeping track of all the pieces. Try labeling the pieces as a, b, c, etc. Then, calculate the volume for each piece. Total all volumes and then calculate the total weight.)

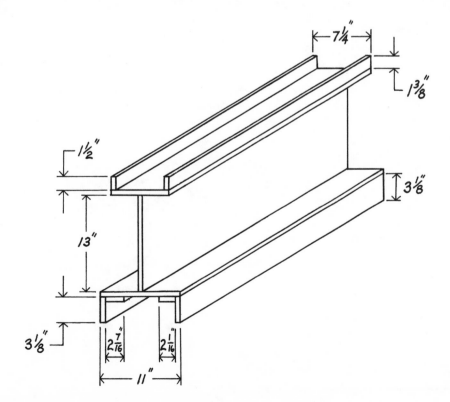

20. Calculate the weight of the following table frame to the nearest tenth of a pound.

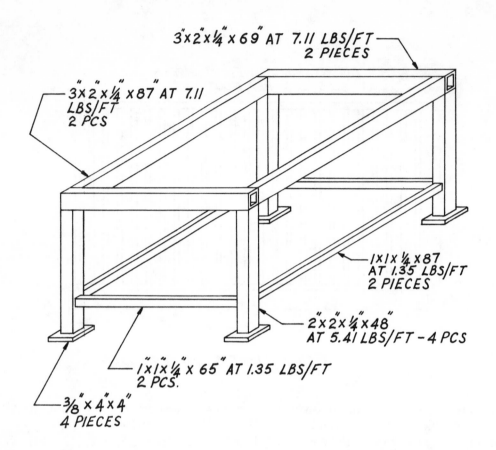

3"x2"x¼"x 69" AT 7.11 LBS/FT
2 PIECES

3"x2"x¼"x87" AT 7.11
LBS/FT
2 PCS

1"x1"x¼"x87"
AT 1.35 LBS/FT
2 PIECES

2"x2"x¼"x48"
AT 5.41 LBS/FT – 4 PCS

1"x1"x¼"x 65" AT 1.35 LBS/FT
2 PCS.

⅜"x 4"x 4"
4 PIECES

BENDING METAL

INTRODUCTION

You will often be called upon to bend metal as part of your job in a welding shop. When a piece of metal is permanently bent, its original flat-out length is changed. Because of this dimensional change, you will have to be able to calculate flat-out lengths of metal objects that have been formed by bending. Most of the bends you will encounter can be grouped into the following types:

1. 90° Sharp Corner Bend.
2. Circle.
3. Half Circle.
4. Quarter Circle.

A short explanation of the effects of bending may help your understanding of the calculations of bending. Why does permanent bending affect the length of a piece of metal? The primary reason is because metal can be stretched and/or compressed. For example, when metal is bent, the thickness of the metal at the bend is compressed on the inside of the corner and stretched on the outside of the corner. The bent piece of metal now has two different lengths; the inside length and the outside length. The required accuracy of the part being produced will determine how accurately these lengths are to be calculated.

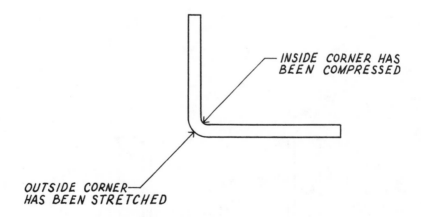

The terms **inside corner** and **outside corner** are clarified by the drawing on the next page. Observer A is looking at the outside corner of the bend. Observer B is looking at the inside corner of the bend.

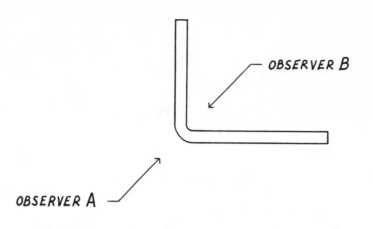

Bending metal can be a complicated manufacturing process involving specialized equipment and formulas. Welders, however, can use a few simple rules to quickly produce excellent, accurately bent parts from flat stock.

90° SHARP CORNER BENDS

It should be pointed out that all bends have a bend radius. Even the so-called 90° sharp corner bend has a bend radius although it is quite small. Because the radius is extremely small, the following rules of thumb assume a bend radius of zero.

OUTSIDE DIMENSIONS

To calculate the flat stock length when given the outside dimensions of the part, subtract twice the metal thickness for each 90° bend.

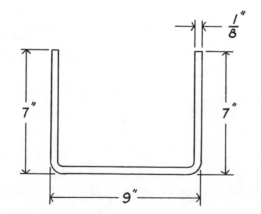

The stretched out length is 22 1/2 in. The flat layout would be as follows:

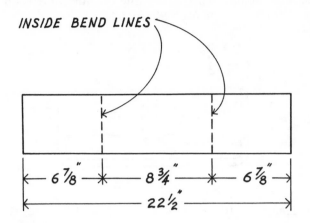

INSIDE DIMENSIONS

To calculate the flat stock length when given the inside dimensions of the part, simply add the inside dimensions. This will give you the flat stock length.

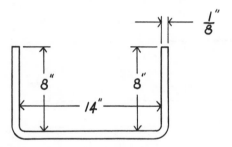

Flat length = 8 + 14 + 8 = 30 in.

The stretched out length is 30 in. The flat layout would be as follows:

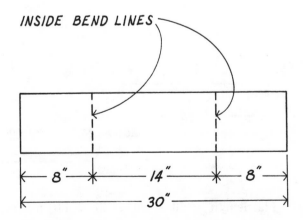

OUTSIDE AND INSIDE DIMENSIONS

To calculate the flat stock length when given one outside dimension and one inside dimension, subtract <u>one</u> metal thickness for each 90° bend.

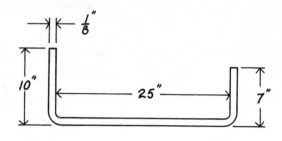

Flat length = 10 + 25 + 7 − 1/8 − 1/8 = 41 3/4 in.

The stretched out length is 41 3/4 in. The flat layout would be as follows:

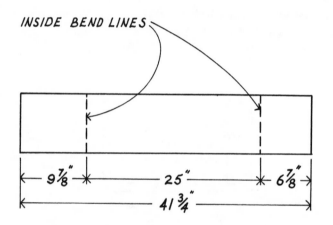

REVIEW OF 90° SHARP CORNER BENDS

The following part has been dimensioned in three different ways. The flat out length is calculated as shown.

INSIDE DIMENSIONS

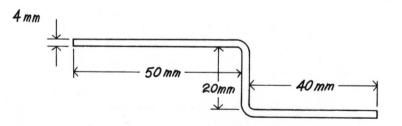

Flat length = 50 + 20 + 40 = 110 mm

OUTSIDE DIMENSIONS

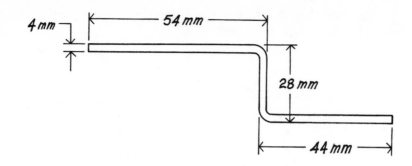

Flat length = 54 + 28 + 44 − 8 − 8 = 110 mm

OUTSIDE AND INSIDE DIMENSIONS

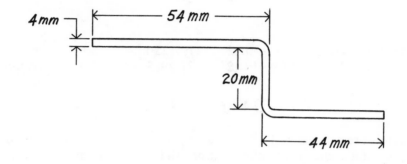

Flat length = 54 + 20 + 44 − 4 − 4 = 110 mm

CIRCLES

A piece of metal bent into a circle has two circumferences. The inside surface has been compressed into a smaller circumference and the outside has been stretched into a larger circumference.

OUTSIDE DIAMETER

To calculate the flat stock length when given the outside diameter, subtract the thickness from the diameter and then calculate the circumference.

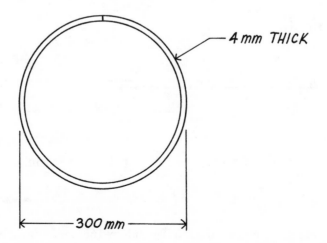

D − thickness = 300 − 4 = 296 mm
$\pi \times 296 = 3.14 \times 296 = 929$ mm (rounded)

INSIDE DIAMETER

To calculate the flat stock length when given the inside diameter, add the thickness to the diameter and then calculate the circumference.

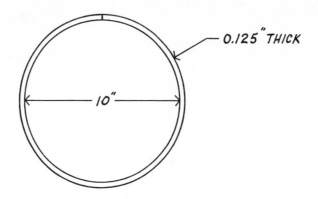

$$D + \text{thickness} = 10 + 0.125 = 10.125 \text{ in.}$$
$$\pi \times 10.125 = 3.14 \times 10.125 = 31.7925 = 31 \ 51/64 \text{ in. (rounded)}$$

HALF CIRCLE

Half circle bends are calculated in the same way as full circle bends. The only difference is that one-half a circumference is calculated rather than a full circumference.

OUTSIDE DIAMETER

To calculate the flat stock length when given the outside diameter, subtract the thickness from the diameter and then calculate one-half of a circumference.

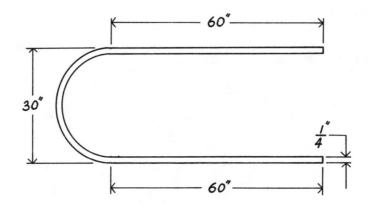

$$D - \text{thickness} = 30 - 1/4 = 29 \ 3/4 \text{ in.}$$

$$\text{HALF CIRCUMFERENCE} = \frac{\pi \times 29 \ 3/4}{2} = \frac{3.14 \times 29.75}{2} = 46 \frac{45}{64} \text{ in. (rounded)}$$

$$\text{Flat length} = 60 + 46 \ 45/64 + 60 = 166 \ 45/64 \text{ in. (rounded)}$$

INSIDE DIAMETER

To calculate the flat stock length when given the inside diameter, add the thickness to the diameter and then calculate one-half of a circumference.

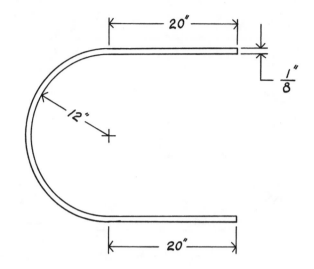

$$D + \text{thickness} = 24 + 1/8 = 24\ 1/8 \text{ in.}$$

$$\text{HALF CIRCUMFERENCE} = \frac{\pi \times 24\ 1/8}{2} = \frac{3.14 \times 24.125}{2} = 37\frac{7}{8} \text{ in. (rounded)}$$

$$\text{Flat length} = 20 + 37\ 7/8 + 20 = 77\ 7/8 \text{ in.}$$

QUARTER CIRCLE

Quarter circle bends are calculated in the same way as full circle bends. The only difference is that one-quarter of a circumference is calculated rather than a full circumference.

INSIDE DIAMETER

To calculate the flat stock length when given the inside diameter, add the thickness to the diameter and then calculate one-quarter of a circumference.

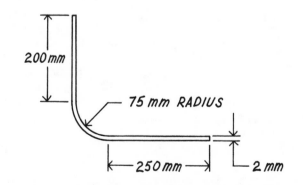

$$D + \text{thickness} = 150 + 2 = 152 \text{ mm}$$

$$\text{QUARTER CIRCUMFERENCE} = \frac{\pi \times 152}{4} = \frac{3.14 \times 152}{4} = 119 \text{ mm}$$

$$\text{Flat length} = 200 + 119 + 250 = 569 \text{ mm}$$

OUTSIDE DIAMETER

To calculate the flat stock length when given the outside diameter, subtract the thickness from the diameter and then calculate one-quarter of a circumference.

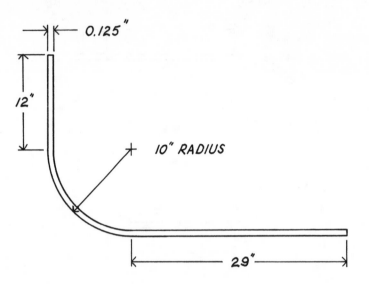

$$D - \text{thickness} = 20 - 0.125 = 19.875 \text{ in.}$$

$$\text{QUARTER CIRCUMFERENCE} = \frac{\pi \times 19.875}{4} = \frac{3.14 \times 19.875}{4} = 15.6 \text{ (rounded)}$$

$$\text{Flat length} = 12 + 15.6 + 29 = 56.6 \text{ in.}$$

KEY TERMS FOR WELDERS
INSIDE CORNER
OUTSIDE CORNER

Unit 22—Practicing Calculations Used in Bending Metal

Show all your work. Be certain the columns line up. Box your answers.

1. Seventeen brackets are required as shown. What is the total length of metal required?

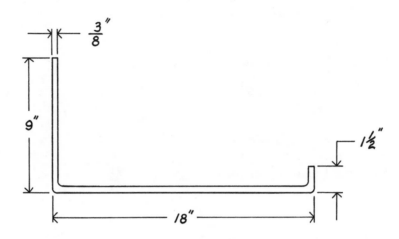

2. Calculate the flat out dimensions of part Ⓐ.

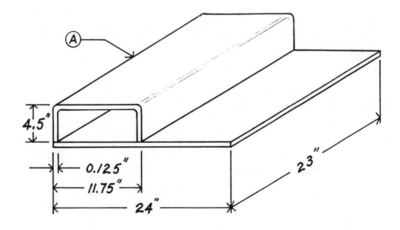

3. Calculate the flat out length.

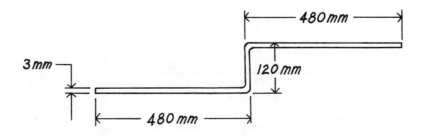

4. What is the flat out length and width of this 1/4 in. thick tank to the nearest thirty-second?

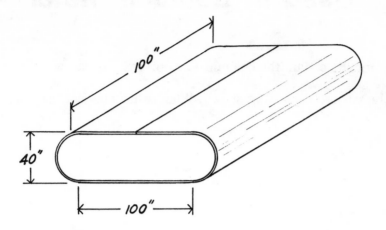

5. Calculate the flat out length of this piece.

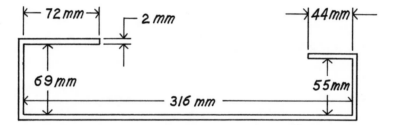

6. Calculate the flat out length of this 3/16 in. thick piece.

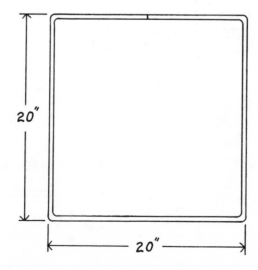

7. Calculate the flat out length of this 3/16 in. thick piece.

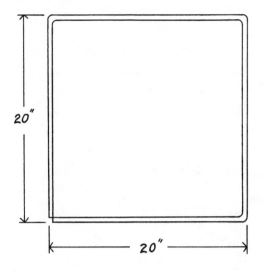

20″

20″

8. Calculate the flat out length of this 3 mm thick piece.

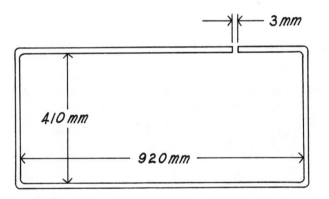

3 mm

410 mm

920 mm

9. Calculate the flat out length of this 1/4 in. thick ring to the nearest thirty-second.

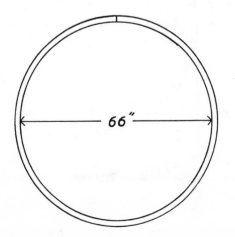

66″

10. Calculate the flat out dimension for this piece to the nearest sixty-fourth.

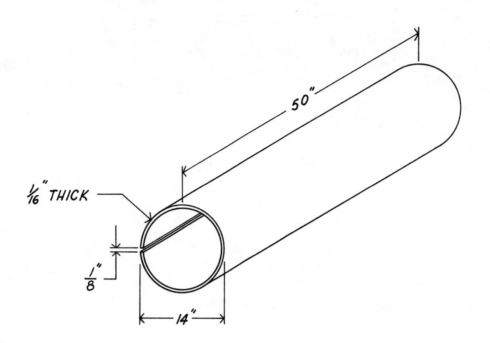

11. What is the flat out length of the following piece to the nearest millimeter?

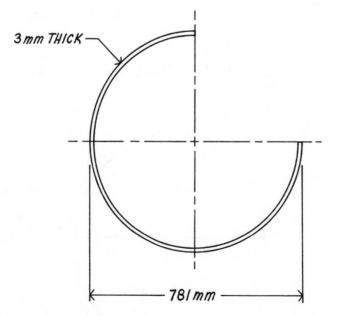

12. Calculate the flat out length of this 2 mm thick piece of steel.

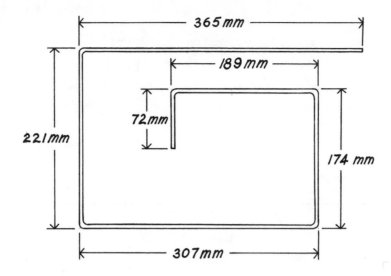

13. Calculate the flat out length. The metal is 1/16 in. thick.

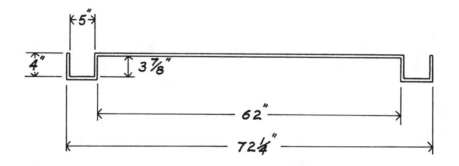

14. Calculate the flat out length to the nearest sixty-fourth.

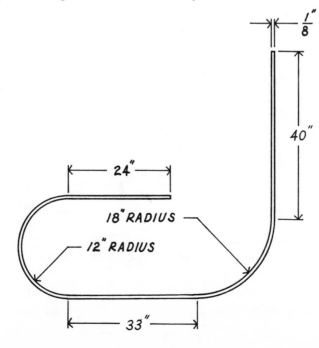

15. Calculate the flat out length.

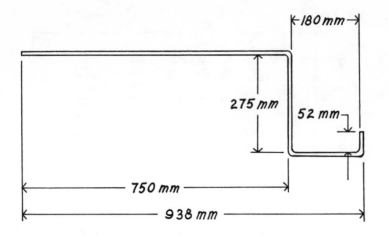

Section Six

PERCENTAGES AND THE METRIC SYSTEM

Objectives for Section Six, Percentages and the Metric System

After studying this section, you will be able to:

- Define percentage
- Convert between a fraction, a decimal, and a percentage
- Calculate percentages
- List the seven basic metric units
- Explain the derived units of measure from length and mass
- List the standard units of metric length
- Convert measurements of length
- Estimate and visualize metric lengths
- Explain derived units of metric length
- List the standard units of metric mass
- Convert measurements of mass
- Estimate and visualize metric mass
- Explain derived units of metric mass
- Demonstrate the proper way to use the Language of SI

Contents for Section Six, Percentages and the Metric System

PERCENTAGES

INTRODUCTION

Percent is a word which expresses things "by the hundred." Percents are parts of the whole of anything when it is divided into 100 equal parts. For example, 80 parts out of 100 can be expressed as 80%. The symbol, %, next to a number tells you it is a percentage number. The use of percents is very common in daily working life. Items such as the following are usually expressed in percentages.

1. The composition of steel.
2. The composition of alloys.
3. Salary deductions.
4. Scrap rate on a job.
5. Quality control statistics.

Being able to solve problems involving percentage is a useful skill. Mathematically, a variety of things can be done with percent. Percents can be converted to decimals or fractions and vice versa. The percent of a number can be calculated. One number can be expressed as a percent of another.

CHANGE A PERCENT TO A FRACTION

Since percents refer to hundredths, changing them to fractions is quite easy. Simply remove the percent sign and write the number as a fraction with a denominator of 100. Reduce if necessary.

$$
\begin{aligned}
75\% &= 75/100 = 3/4 \\
100\% &= 100/100 = 1 \\
125\% &= 125/100 = 1\ 25/100 = 1\ 1/4 \\
3000\% &= 3000/100 = 30
\end{aligned}
$$

The above method works fine for most cases, but if the percent is a fraction or a mixed number, there is an additional step. First, change the mixed number to an improper fraction. Now that you have the percent in fractional form, remove the percent sign, multiply the denominator by 100 and then reduce, if necessary.

$$
\begin{aligned}
1/4\% &= 1/400 \\
3/16\% &= 3/1600 \\
12\ 1/2\% &= 25/2\% = 25/200 = 1/8 \\
107\ 1/2\% &= 215/2\% = 215/200 = 1\ 15/200 = 1\ 3/40
\end{aligned}
$$

CHANGE A PERCENT TO A DECIMAL

This is done using the same method as explained for changing percents to fractions, except the final answer is expressed as a decimal number. Remove the percent sign, divide the number by 100, and then express the answer as a decimal.

$$67\% = 67/100 = 0.67$$
$$16.6\% = 16.6/100 = 0.166$$
$$219\% = 219/100 = 2.19$$
$$1019\% = 1019/100 = 10.19$$

In cases where the percent is a fraction or a mixed number, first change the fraction to a decimal.

$$1/16\% = 0.0625\% = 0.0625/100 = 0.000625$$
$$11\ 1/8\% = 11.125\% = 11.125/100 = 0.11125$$
$$95\ 1/2\% = 95.5\% = 95.5/100 = 0.955$$
$$365\ 1/4\% = 365.25\% = 365.25/100 = 3.6525$$

CHANGE A FRACTION TO A PERCENT

The easiest method of changing fractions, mixed numbers, and whole numbers to percents is to multiply by 100 and add a percent sign.

$$9/10 = 9/10 \times 100 = 90\%$$
$$1 = 1 \times 100 = 100\%$$
$$2\ 3/4 = 11/4 \times 100 = 275\%$$
$$5\ 3/7 = 38/7 \times 100 = 542\ 6/7\%$$

CHANGE A DECIMAL TO A PERCENT

To change decimals to percents, multiply by 100 and add a percent sign.

$$0.01 = 0.01 \times 100 = 1\%$$
$$0.25 = 0.25 \times 100 = 25\%$$
$$1.0 = 1.0 \times 100 = 100\%$$
$$3.5 = 3.5 \times 100 = 350\%$$

CALCULATE A PERCENT OF A NUMBER

This type of question is usually expressed as "find 25% of 300." You may recall from Unit 9, Multiplication of Fractions, that "of" means to multiply. Therefore, the solution is arrived at by multiplying 25% × 300. To do this, change the percent to a decimal and proceed with the multiplication. In this example, 0.25 × 300 = 75. 25 percent of 300 is 75.

The percent can also be calculated in fractional form rather than as a decimal. This is especially useful if the decimal form is a repeating decimal such as 0.666 . . .

$$30\% \text{ of } 3750 = 0.30 \times 3750 = 1125$$
$$112\% \text{ of } 43 = 1.12 \times 43 = 48.16$$
$$1.5\% \text{ of } 900 = 0.015 \times 900 = 13.5$$
$$16\ 2/3\% \text{ of } 2700 = 50/3\% \text{ of } 2700 = 50/300 \times 2700 = 450$$

CALCULATE THE PERCENTAGE ONE NUMBER IS OF ANOTHER NUMBER

This type of question is usually expressed as, "what percent of 52 is 13?" The word "of" means, to multiply, and "is" means, equals. Therefore, the questions can be expressed as ?% × 52 = 13. To solve this problem, form a fraction with the multiplier, 52, as the denominator and, 13, as the numerator.

$$?\% \times 52 = 13$$
$$?\% = 13/52$$

Now, express this fraction as a percent.

$$13/52 \times 100 = 1300/52 = 25 \text{ percent}$$
13 is 25 percent of 52

The way in which this type of question is expressed may cause some confusion. For example, the question was originally expressed as "what percent of 52 is 13?" It could also have been expressed as, "13 is what percent of 52?" Since "is" means equals and "of" means times, the resulting equation is 13 = ?% × 52. Both expressions have the same meaning.

KEY TERMS FOR WELDERS
PERCENT
%

Unit 23—Practicing Percentages

Show all your work. Be certain the columns line up. Box your answers.

1. Express the following percents as fractions.

(a) 12%

(b) 0.1%

(c) 0.05%

(d) 1/16%

(e) 1010%

(f) 5/7%

2. Express the following percents as decimals.

    (a) 75%   (b) 29%

    (c) 14.14%   (d) 0.125%

    (e) 0.009%   (f) 2/5%

3. Express the following fractions as percents.

    (a) 1/4   (b) 1/100

    (c) 33 1/3   (d) 2560 1/2

    (e) 11/19   (f) 99 1/16

4. Express the following decimals as percents.

    (a) 0.125   (b) 1.125

    (c) 0.017   (d) 140.3

    (e) 0.002   (f) 0.98

5. Calculate the following to two decimal places.

    (a) 45% of 823   (b) 9.4% of 507

    (c) 100% of 28.5   (d) 33 1/3% of 253

    (e) 0.17% of 123.6   (f) 1938% of 22

6. Calculate the following to two decimal places.

   (a) What percent of 432 is 95?

   (b) 579 is what percent of 606?

   (c) 1 is what percent of 2?

   (d) What percent of 15 is 139?

   (e) What percent of 50 is 225?

   (f) 31.2 is what percent of 65.7?

7. A welder's suggestion to use a different type of fixture resulted in a 5% increase in production. The previous production rate was 180 parts per hour. What is the new production rate?

8. In 1985 American mills delivered 68 million tons of steel. The following year shipments increased by 8.5%. How many tons of steel were delivered in 1986?

9. Two hundred and eighteen motorcycles started in the Illinois-Michigan Motorcycle Rally. Only eighty-five bikes finished the rally. What percent of the bikes did not finish? Calculate to the nearest percent.

10. A welding shop was allowed a discount of 12 1/2% on the regular price of $426.00 for a pedestal grinder. Calculate the price the shop paid to the nearest cent.

11. The Metal Building Association reported that sales this year were 31 1/2% ahead of last year's total of $1,266,000,000. What was the dollar value increase in metal building sales this year?

12. The following employees received a salary increase of 5.25%. Calculate the weekly gross pay, income tax, union dues, and social security to the nearest cent.

	WEEKLY GROSS PAY		INCOME TAX		UNION DUES	SOCIAL SECURITY
	Before Increase	After Increase	%	$	3 1/2% of Gross Pay	4% of Gross Pay
Calleja, Nick	453.60		16%			
Ing, Richard	491.60		18%			
Lima, Jose	354.00		15%			
Massey, Vincent	320.00		15%			
Shulman, Lorna	540.80		19%			
Simpson, Katie	402.00		18%			
Williams, Deroy	458.80		15%			
Zakoor, Eli	398.80		17%			

THE METRIC SYSTEM

TOWARD A METRIC SOCIETY

Anyone who has been involved with the metal working trades realizes that the global market is a powerful force leading us toward a metric society. The decision by major manufacturers to change to metric has guaranteed that it will be the dominant system in the workplace.

The welding industry is just one of many industries in the process of metric conversion. As a welder, you will be expected to understand and work in metric. You will be pleased to know it is not a difficult system to learn. It is a very logical, very consistent system. However, the biggest difficulty most people have is with changing habits. We are so accustomed to using the inch/pound units, that giving them up can be quite difficult. But, as job requirements begin to include "the ability to work in metric," an attachment to the inch/pound system usually gives way to the reality that an understanding of metrics is an important part of your training as a welder.

THE METRIC SYSTEM

The metric system, since it is a measuring system, is used for all types of measurement. In fact, the metric system (also referred to as SI; see the Glossary) is intended to be a system of measurement for everything measurable! The creation of SI was a worldwide cooperative undertaking. Committees of experts were formed to find the best possible way of measuring. All this effort resulted in the establishment of what is known as the seven basic units of SI. The outstanding characteristic of the seven basic units is that any physical quantity in the universe can be measured by one of the seven basic units.

THE BASIC UNITS

The seven basic units are:

1. Length.
2. Mass.
3. Time.
4. Electrical current.
5. Temperature.
6. Luminous intensity
7. Substance

(Note: There are also a few supplementary units, see the Glossary.)

In welding, you will be required to know and apply the two basic units of length and mass. In everyday use, the word "weight" is often used to mean mass, although they have different meanings (see the Glossary). In this text, weight and mass will be considered to have the same meaning.

DERIVED UNITS OF MEASURE

The basic units are subdivided into parts called derived units of measure. For example, the basic units of length and mass have the following derived units.

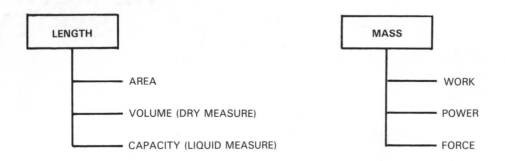

It is usually not necessary for a welder to be completely familiar with the derived units of mass.

LENGTH

Length is one of the basic units you will work with in the metric world. The originators of SI decided each unit of measure should have a base unit as a reference point. Then, subdivisions and multiples of the base unit were agreed upon and named. Their lengths and names were all related to the base unit, making a very consistent system. The meter was chosen as the base unit for length.

STANDARDIZED UNITS OF LENGTH

The following chart shows the base unit (the meter) and some of the subdivisions and multiples established. The very large multiples and the very small subdivisions have been omitted from the chart in order to focus on those most common in ordinary daily life.

NAME	SYMBOL	HOW EACH UNIT OF LENGTH IS RELATED TO THE BASE UNIT (METER)
Kilometer	km	1000 meters
Hectometer	hm	100 meters
Dekameter	dam	10 meters
Meter	m	1 meter
Decimeter	dm	0.1 meter
Centimeter	cm	0.01 meter
Millimeter	mm	0.001 meter

BASE UNIT → Meter

Start reading the chart at the base unit (the meter) and work your way up and down from there. Do not be concerned about what these lengths actually look like. You will get familiar with this later. For now, just observe how well the system is organized. The names of all units are related by their suffixes. They all end in "meter." Even the abbreviations are related; they all end in "m." Also, all units are related to the base unit in multiples of 10. In welding, you will not be working with the hectometer, dekameter, and decimeter. They are illustrated here so that you can see the regularity and pattern of the system.

CONVERTING ONE UNIT OF MEASURE TO ANOTHER

The following diagram is a useful device for learning how to convert measurements.

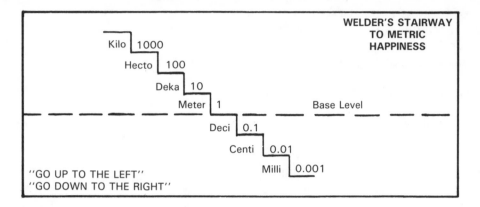

1. Convert 1 629 m to km.
    Looking at the stairway, you have to go up three steps (or places) to the left to move from m to km. So, in this case, you would move the decimal point three places to the left. Therefore, 1 629 m equals 1.629 km.
2. Convert 85 cm to mm.
    Looking at the stairway, you have to go down one step to the right to move from cm to mm. In this case, you would move the decimal one place to the right. Therefore, 85 cm equals 850 mm.
3. Convert 7.5 km to dm.
    Go down the stairway four places to the right. Therefore, the decimal moves four places to the right; 7.5 km equals 75 000 dm.

The "metric stairway" is a useful visual tool for converting measurements. There is, of course, a mathematical explanation why conversion from one unit to another can be done simply by moving the decimal. Since all units are related in multiples of ten, converting consists of multiplying by these multiples and, in effect, moving the decimal.

4. Convert 5.9 km to m.
    $5.9 \times 1\ 000 = 5\ 900$ m.
5. Convert 16 cm to mm.
    $16 \times 10 = 160$ mm.
6. Convert 1 437 dm to hm.
    $1\ 437 \times 0.001 = 1.437$ hm.

VISUALIZING METRIC LENGTHS

Becoming knowledgeable in the metric system requires that you develop an awareness of the actual lengths of metric units. Just as you already have a rough idea of the length of one inch, one foot, and one yard, you can also start becoming aware of metric lengths. In order to avoid being overwhelmed with all the units of length, it is best just to concentrate on the millimeter, centimeter, and meter. Being familiar with these three will satisfy your needs in welding. In fact, almost all the metric blueprints you will be reading are dimensioned only in millimeters. Below is a full size metric scale 150 mm long. If you do not own a metric scale, you should buy one now.

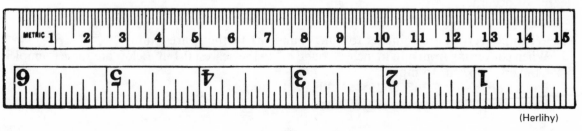

(Herlihy)

Here are a few examples to help you visualize the millimeter, centimeter, and meter.

- A dime is about 1 mm thick.
- A thick line drawn with a dull pencil is about 1 mm wide.
- A slice of bread is about 1 cm thick.
- A stack of ten dimes is about 1 cm high.
- One meter is about three inches longer than one yard.
- The height of a door knob is about one meter from the floor.

ESTIMATING METRIC LENGTHS

A useful technique is to memorize certain body dimensions, then use these to estimate other dimensions.

Measure the following to the nearest mm:
    The width of your thumb in mm;

    The width of your palm in mm;

    The span of your hand from the smallest finger to thumb in cm;

    The distance from your elbow to your fingertips in cm;

    The length of your stride in cm;

    Your height in m.

Estimate, using the span of your hand as a guide, the length of:
    An object and the number of times it takes to span the length with your hand.

    Multiply the number of spans by the length of a span to arrive at an estimate of the length. If your hand has a span of 24 cm and an object is 5 spans long, its estimated length is 24 times 5 or 120 cm.

    Check your accuracy with a metric scale.

CONVERTING METRIC AND CUSTOMARY

At some time in your job, you will probably have to make conversions between the two systems. Conversion charts or cards are often available, but you should know the mathematical method for converting. Your main concern as a welder will be converting from mm to inches and from inches to mm. Therefore, these are the only conversions that will be explained here.

Calculations in converting often lead to the necessity of rounding the answer. Answers can be rounded for maximum mathematical accuracy or they can be rounded for the accuracy normally required in the shop. A method for shop accurate conversions will be explained here.

SHOP ACCEPTABLE ACCURACY

In general, it can be stated that the closest tolerance to which a welder may be expected to work is to the nearest 1/64 in. The following guidelines ensure that your work stays within this degree of accuracy.

1. When converting from inches to mm, round off to the nearest mm.
2. When converting from mm to inches, round off to the nearest 0.1 in. Below are some examples:
   Convert 7.367 in. to mm.
   Exact conversion $7.367 \times 25.4 = 187.1\,218$ mm.
   Shop acceptable accuracy = 187 mm.

Convert 11 1/16 in. to mm.
Exact conversion 11.0 625 × 25.4 = 280.9 875 mm.
Shop acceptable accuracy 281 mm.

Convert 65.5 mm to inches.
Exact conversion 65.5 ÷ 25.4 = 2.5787402 in.
Shop acceptable accuracy 2.6 in.

Convert 485.35 mm to inches.
Exact conversion 485.35 ÷ 25.4 = 19.108268 in.
Shop acceptable accuracy 19 in.

Two rules for converting are stated below:
1. When converting from inches to millimeters, multiply by 25.4.
2. When converting from millimeters to inches, divide by 25.4.

DERIVED UNITS OF LENGTH

As mentioned earlier, the basic unit of length has three derived units of length: area, volume, and capacity.

AREA

Area is calculated in the normal manner, depending upon the shape of the object. Results are expressed as mm^2, km^2, etc. Be sure you are calculating with the same units; that is, mm × mm or cm × cm. Do not, for example, multiply meters times millimeters to calculate an area.

VOLUME

Volume refers to dry measure and is calculated in the normal manner, depending upon the shape of the object. Results are expressed as m^3, mm^3, etc. As pointed out with area, be certain you are calculating with the same units: $m × m × m = m^3$.

CAPACITY

Capacity is a derived unit of length used for liquid measure. The base unit is the liter. The standardized units of capacity are illustrated in the chart below.

NAME	SYMBOL	HOW EACH UNIT OF CAPACITY IS RELATED TO THE BASE UNIT (LITER)
Kiloliter	kl	1 000 liters
Hectoliter	hl	100 liters
Dekaliter	dal	10 liters
Liter (BASE UNIT)	L	1.0 liter
Deciliter	dl	0.1 liter
Centiliter	cl	0.01 liter
Milliliter	ml	0.001 liter

As you saw with the chart for units of length, there is a regular pattern of units of capacity. Also, as with units of length, you will work with only a few of the units of capacity such as the liter (L) and milliliter (ml). The other units are illustrated here so you can see the pattern of the system.

CONVERTING ONE UNIT OF MEASURE TO ANOTHER

To convert from one unit of measure to another, use the following diagram: (Examples follow.)

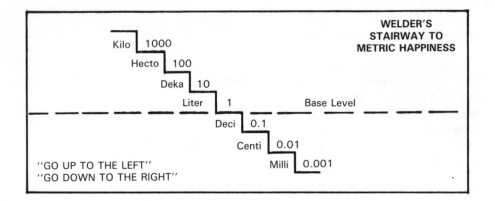

Convert 49 liters to milliliters.
49 liters = 49 000 milliliters.

Convert 7 kiloliters to liters.
7 kiloliters = 7 000 liters.

Convert 28 milliliters to liters.
28 milliliters = 0.028 liters.

CONVERTING METRIC AND CUSTOMARY

To convert between metric and standard capacity, use the conversion figure of one gallon = 3.785 liters.

VISUALIZING METRIC CAPACITY

Here are a few examples to help you visualize the liter and milliliter.
• A teaspoon is about 5 ml.
• A bowl of soup is about 200 ml.
• The liter is approximately the volume of four paper coffee cups.
• A tea kettle has a capacity of about 2 liters.

MASS

Mass is one of the basic units you will work with in the metric system. The base unit for mass is the gram. The following chart shows the base unit and some of the sub-divisions and multiples established. The very large multiples and the very small sub-divisions have been omitted in order to focus on those most common in ordinary daily life.

	NAME	SYMBOL	HOW EACH UNIT OF MASS IS RELATED TO THE BASE UNIT (GRAM)
	Megagram	Mg (t)	1 000 000 grams
	—	—	—
	—	—	—
	Kilogram	kg	1 000 grams
	Hectogram	hg	100 grams
	Dekagram	dag	10 grams
BASE UNIT →	Gram	g	1 gram
	Decigram	dg	0.1 gram
	Centigram	cg	0.01 gram
	Milligram	mg	0.001 gram

Start reading the chart at the base unit and work your way up and down from there. Observe how well the system is organized. All units are related to the base unit in multiples of 10. The names of all units are related by their suffixes; they all end in "-gram." Even the abbreviations are related; they all end in "-g." There is one variation accepted here. The megagram is also commonly referred to as a metric ton and has the symbol "t."

Of the eight units of mass shown on the chart, you will probably encounter only the megagram (metric ton), kilogram, gram, and milligram. The other units are illustrated here so that you can see the pattern of the system. Since the units increase in multiples of 10, you may have expected a metric unit for 10 000 grams and 100 000 grams. However, there are no metric units between the kilogram (1 000 grams) and the metric ton (1 000 000 grams).

CONVERTING ONE UNIT OF MEASURE TO ANOTHER

The following diagram is a useful device for learning how to convert measurements. (Examples follow.)

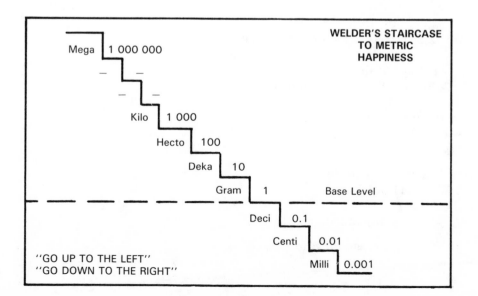

WELDER'S STAIRCASE TO METRIC HAPPINESS

Mega | 1 000 000

Kilo | 1 000

Hecto | 100

Deka | 10

Gram | 1 — Base Level

Deci | 0.1

Centi | 0.01

Milli | 0.001

"GO UP TO THE LEFT"
"GO DOWN TO THE RIGHT"

Convert 3 275 g to kg.
You have to go up three steps (or places) to the left to move from gram to kilograms. So, you would move the decimal point three places to the left. Therefore, 3 275 grams equals 3.275 kilograms.

Convert 987 000 milligrams to kilograms.
You have to go up six places to the left to move from milligrams to kilograms. Move the decimal six places to the left. Therefore, 987 000 milligrams equals 0.987 kilograms.

Convert 48 megagrams to kilograms.
You have to move down three places to the right to move from megagrams to kilograms. Move the decimal three places to the right. Therefore 48 megagrams equals 48 000 kilograms.

As you can see from these and previous examples, converting between units in metric can be done simply by moving the decimal point.

VISUALIZING METRIC MASS

The metric weights you should become familiar with are gram, kilogram, and metric ton. Below are some examples:
• A paper clip has a mass of about 1 g.
• A thumb tack weighs about one gram.
• The mass of about five average sized apples is about 1 kg.
• A heavyweight boxer weighs about 91 kg.
• A compact car has a mass of about 1 metric ton.
• One metric ton weighs about 200 lbs. more than one standard ton.

CONVERTING METRIC AND CUSTOMARY

$$1 \text{ lb.} = 0.4\ 536 \text{ kg.}$$
$$1 \text{ kg.} = 2.205 \text{ lbs.}$$
$$1 \text{ metric ton} = 2{,}205 \text{ lbs.}$$

The above conversions have been rounded and therefore slight inaccuracies will result. They are accurate enough, however, for the work you will encounter in the shop. Below are some examples.

Convert 875 lbs. to kg.
$875 \times 0.4\ 536 = 396.9$ lbs.

Convert 32,000 lbs. to metric tons.
$32{,}000 \div 2{,}205 = 14.5$ metric tons (rounded).

Convert 81.8 kg. to lbs.
$81.8 \times 2.205 = 180.4$ lbs. (rounded).

Convert 100 000 kg. to standard tons.
$$\frac{100\ 000 \times 2.205}{2{,}000} = 110.25 \text{ tons}$$

Convert 5 metric tons to lbs.
$5 \times 2{,}205 = 11{,}025$ lbs.

Convert 30 metric tons to standard tons.
$$\frac{30 \times 2{,}205}{2{,}000} = 33 \text{ standard tons (rounded).}$$

THE LANGUAGE OF SI

1. Never use a period after a symbol unless it is at the end of a sentence.
   Example: Her height is 148 cm but his is 185 cm.
2. Symbols are never made plural.
   Example: 1 g   10 g   100 g   1 000 g
3. Symbols are almost always written in lower case. The two exceptions so far are megagram and liter.
   Example: m for meter, g for gram, but Mg for megagram and L and liter.
4. Never start a sentence with a symbol. Write the name out in full.
   Example: Kilometers measure distance.
5. Always leave a space between the quantity and the symbol.
   Example: 15 km   6 m   24 g
6. If there is no quantity with the unit, spell the unit out in full.
   Example: Meters are marvelous and grams are great.
7. Always use decimals, not fractions.
   Example: 0.25 g   1.5   2.75 m
8. Always place a zero before the decimal point if the value is less than one.
   Example: 0.125 L   0.75 kg   0.5 m
9. Use spaces, not commas, to separate large numbers into easily readable three digit blocks.
   Example: 15 764   5.7 642   123 456

Unit 24—Practicing with the Metric System

Show all your work. Be certain the columns line up. Box your answers.
1. (a) Name the seven basic units in the metric system.

   (b) Which two of the seven basic units will you, as a welder, encounter?

2. Name the base unit for length.

3. How many millimeters are there in one meter?

4. How many centimeters are there in one meter?

5. Convert the following:
   (a) 2.5 kilometers to millimeters.

   (b) 1 795 centimeters to meters.

   (c) 12 800 millimeters to kilometers.

(d) 2 985 400 meters to kilometers.

(e) 1 000 centimeters to millimeters.

(f) 595 decimeters to kilometers.

6. Convert the following to the nearest millimeter:
   (a) 0.125 in. to mm.

   (b) 60 in. to mm.

   (c) 5 ft. 8 in. to mm.

   (d) 16 1/2 in. to mm.

   (e) 30 ft. to m.

   (f) 100 yards to m.

7. Convert the following to the nearest 1/64 in.:
   (a) 1 000 mm to inches.

   (b) 5.8 mm to inches.

   (c) 114.5 mm to inches.

   (d) 1 000 m to feet and inches.

   (e) 3 855 mm to feet and inches.

   (f) 10 km to feet and inches.

8. Name the base unit for liquid capacity.

9. How many milliliters are there in a liter?

10. Convert the following:
    (a) 3 785 L to gallons.

    (b) 128 ml to L.

(c) 45 gallons to L.

(d) 6.7 L to ml.

(e) 1 gallon to ml.

(f) 10 000 ml to gallons (to nearest hundredth).

11. Name the base unit for mass.

12. How many grams are there in one kilogram?

13. Convert the following:
    (a) 77 590 g to mg.

    (b) 39.25 kg to g.

    (c) 0.87 kg to mg.

    (d) 222 mg to g.

    (e) 1 000 mg to kg.

    (f) 44 g to kg.

14. Convert the following:
    (a) 1,000 lbs. to kg.

    (b) 6 472 g to lbs. (to nearest hundredth).

    (c) 95 lbs. to g.

    (d) 3.5 standard tons to kg.

    (e) 4.25 t to standard tons (to nearest hundredth).

    (f) 75 standard tons to kg.

USEFUL
INFORMATION

FRACTION, DECIMAL, AND MILLIMETER CONVERSIONS

FRACTIONAL INCH	DECIMAL INCH	MILLIMETER	FRACTIONAL INCH	DECIMAL INCH	MILLIMETER
1/64	.015625	0.397	33/64	.515625	13.097
1/32	.03125	0.794	17/32	.53125	13.494
3/64	.046875	1.191	35/64	.546875	13.891
1/16	.0625	1.588	9/16	.5625	14.288
5/64	.078125	1.984	37/64	.578125	14.684
3/32	.09375	2.381	19/32	.59375	15.081
7/64	.109375	2.778	39/64	.609375	15.478
1/8	.125	3.175	5/8	.625	15.875
9/64	.140625	3.572	41/64	.640625	16.272
5/32	.15625	3.969	21/32	.65625	16.669
11/64	.171875	4.366	43/64	.671875	17.066
3/16	.1875	4.762	11/16	.6875	17.462
13/64	.203125	5.159	45/64	.703125	17.859
7/32	.21875	5.556	23/32	.71875	18.256
15/64	.234375	5.953	47/64	.734375	18.653
1/4	.25	6.350	3/4	.75	19.05
17/64	.265625	6.747	49/64	.765625	19.447
9/32	.28125	7.144	25/32	.78125	19.844
19/64	.296875	7.541	51/64	.796875	20.241
5/16	.3125	7.938	13/16	.8125	20.638
21/64	.328125	8.334	53/64	.828125	21.034
11/32	.34375	8.731	27/32	.84375	21.431
23/64	.359375	9.128	55/64	.859375	21.828
3/8	.375	9.525	7/8	.875	22.225
25/64	.390625	9.922	57/64	.890625	22.622
13/32	.40625	10.319	29/32	.90625	23.019
27/64	.421875	10.716	59/64	.921875	23.416
7/16	.4375	11.112	15/16	.9375	23.812
29/64	.453125	11.509	61/64	.953125	24.209
15/32	.46875	11.906	31/32	.96875	24.606
31/64	.484375	12.303	63/64	.984375	25.003
1/2	.5	12.700	1	1.	25.400

USEFUL CONVERSIONS BETWEEN THE METRIC AND CONVENTIONAL SYSTEMS

LENGTH

1 inch	= 25.4 mm	1 mm	= 0.03937 inches
1 inch	= 2.54 cm	1 cm	= 0.3937 inches
1 foot	= 30.48 cm	1 cm	= 0.0328 feet
1 foot	= 0.3048 m	1 m	= 3.28 feet
1 yard	= 0.9144 m	1 m	= 1.0936 yard
1 mile	= 1.609 km	1 km	= 0.621 mile

AREA

1 in.^{2}	= 645.2 mm^2	1 mm^2	= 0.00155 in.^{2}
1 in.^{2}	= 6.452 cm^2	1 cm^2	= 0.155 in.^{2}
1 ft.^{2}	= 929.03 cm^2	1 cm^2	= 0.00108 ft.^{2}
1 ft.^{2}	= 0.093 m^2	1 m^2	= 10.764 ft.^{2}
1 yd.^{2}	= 0.836 m^2	1 m^2	= 1.196 yd.^{2}
1 square mile	= 2.59 km^2	1 km^2	= 0.386 square mile

VOLUME (DRY)

1 in.^{3}	= 16.388 cm^3	1 cm^3	= 0.061 in.^{3}
1 ft.^{3}	= 0.028 m^3	1 m^3	= 35.31 ft.^{3}
1 yd.^{3}	= 0.765 m^3	1 m^3	= 1.308 yd.^{3}

CAPACITY (FLUID)

U.S.		**Canadian**
1 quart	= 0.946 L	1.14 L
1 gallon	= 3.785 L	4.546 L
1 L	= 1.057 quarts	0.877 quarts
1 L	= 0.264 gallons	0.220 gallons

WEIGHT

1 oz.	= 28.35 g	1 g	= 0.0353 oz.
1 lb.	= 453.59 g	1 kg	= 2.205 lb.
1 lb.	= 0.4536 kg	1 ton (metric)	= 2,204.6 lb.

NOTE: Converting between the Standard and Metric systems often results in highly unwieldy numbers. Because of this, many of the above conversions have been rounded, but are quite accurate as a quick reference.

BENDING METAL
(How to Calculate Flat Stock Lengths Assuming a Zero Bend Radius)

HALF CIRCLE

1. Given the outside diameter, subtract the thickness from the diameter and then calculate one-half a circumference; or
2. Given the inside diameter, add the thickness to the diameter and then calculate one-half a circumference.

QUARTER CIRCLE

1. Given the outside diameter, subtract the thickness from the diameter and then calculate one-quarter of a circumference; or
2. Given the inside diameter, add the thickness to the diameter and then calculate one-quarter of a circumference.

90° CORNER BEND

1. Given the outside dimensions, add the outside dimensions and subtract twice the metal thickness for each 90° bend; or
2. Given the inside dimensions, add the inside dimensions; or
3. Given an outside and an inside dimension, add the dimensions and subtract one metal thickness for each 90° bend.

CIRCLE

1. Given the outside diameter, subtract the thickness from the diameter and then calculate the circumference; or
2. Given the inside diameter, add the thickness to the diameter and then calculate the circumference.

INCHES CONVERTED TO DECIMALS OF FEET

Inches	Decimal of a Foot	Inches	Decimal of a Foot	Inches	Decimal Of a Foot
1/8	.01042	3 1/8	.26042	6 1/4	.52083
1/4	.02083	3 1/4	.27083	6 1/2	.54167
3/8	.03125	3 3/8	.28125	6 3/4	.56250
1/2	.04167	3 1/2	.29167	7	.58333
5/8	.05208	3 5/8	.30208	7 1/4	.60417
3/4	.06250	3 3/4	.31250	7 1/2	.62500
7/8	.07291	3 7/8	.32292	7 3/4	.64583
1	.08333	4	.33333	8	.66666
1 1/8	.09375	4 1/8	.34375	8 1/4	.68750
1 1/4	.10417	4 1/4	.35417	8 1/2	.70833
1 3/8	.11458	4 3/8	.36458	8 3/4	.72917
1 1/2	.12500	4 1/2	.37500	9	.75000
1 5/8	.13542	4 5/8	.38542	9 1/4	.77083
1 3/4	.14583	4 3/4	.39583	9 1/2	.79167
1 7/8	.15625	4 7/8	.40625	9 3/4	.81250
2	.16666	5	.41667	10	.83333
2 1/8	.17708	5 1/8	.42708	10 1/4	.85417
2 1/4	.18750	5 1/4	.43750	10 1/2	.87500
2 3/8	.19792	5 3/8	.44792	10 3/4	.89583
2 1/2	.20833	5 1/2	.45833	11	.91667
2 5/8	.21875	5 5/8	.46875	11 1/4	.93750
2 3/4	.22917	5 3/4	.47917	11 1/2	.95833
2 7/8	.23959	5 7/8	.48958	11 3/4	.97917
3	.25000	6	.50000	12	1.00000

SUMMARY OF FORMULAS

PERIMETER OF A SQUARE

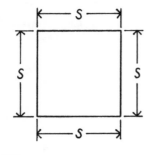

P = Perimeter
S = Side

P = S + S + S + S

PERIMETER OF A RECTANGLE

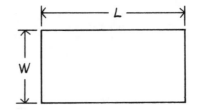

P = Perimeter
L = Length
W = Width

P = L + L + W + W

AREA OF A SQUARE

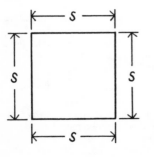

A = Area
S = Side

A = S × S

AREA OF A RECTANGLE

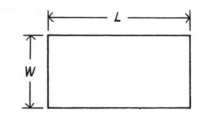

A = Area
L = Length
W = Width

A = L × W

PERIMETER OF A PARALLELOGRAM

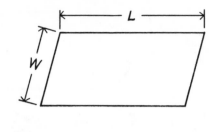

P = Perimeter
L = Length
W = Width

P = L + L + W + W

PERIMETER OF A TRIANGLE

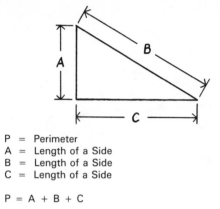

P = Perimeter
A = Length of a Side
B = Length of a Side
C = Length of a Side

P = A + B + C

AREA OF A PARALLELLOGRAM

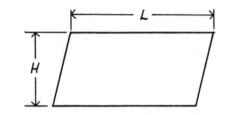

A = Area
L = Length
H = Height

A = L × H

AREA OF A TRIANGLE

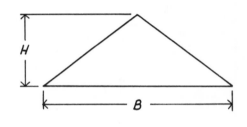

A = Area
B = Length of the Base
H = Height

A = 1/2 × B × H

PERIMETER OF A TRAPEZOID

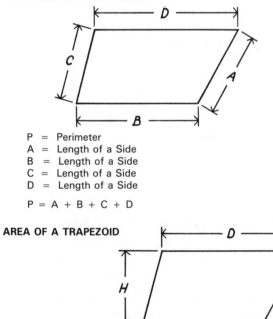

P = Perimeter
A = Length of a Side
B = Length of a Side
C = Length of a Side
D = Length of a Side

P = A + B + C + D

CIRCUMFERENCE OF A CIRCLE

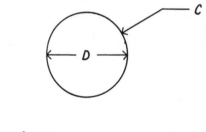

C = Circumference
D = Diameter
π = 3.14

C = π × D

AREA OF A TRAPEZOID

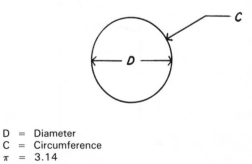

A = Area
B = Length of a Parallel Side
D = Length of a Parellel Side
H = Height

A = 1/2 × (B + D) × H

Note: To use this formula, first add the two figures inside
the brackets, then perform the multiplication.

DIAMETER OF A CIRCLE

D = Diameter
C = Circumference
π = 3.14

$$D = \frac{C}{\pi}$$

AREA OF A CIRCLE

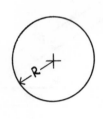

A = Area
R = Radius
π = 3.14

A = π × R × R

PERIMETER OF A HALF CIRCLE

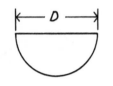

P = Perimeter
D = Diameter
π = 3.14

$$P = (\frac{\pi \times D}{2}) + D$$

Note: To use this formula, first perform the calculations inside the brackets, then add the result to D.

AREA OF A HALF CIRCLE

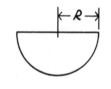

A = Area
R = Radius
π = 3.14

$$A = \frac{\pi \times R \times R}{2}$$

Note: To use this formula, first perform the multiplication, then divide the result by 2.

AREA OF THE CURVED SURFACE OF A CYLINDER

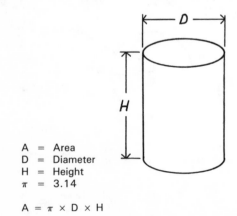

A = Area
D = Diameter
H = Height
π = 3.14

A = π × D × H

PERIMETER OF A SEMI-CIRCULAR SIDED SHAPE

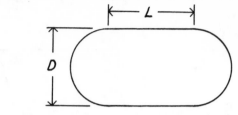

P = Perimeter
D = Diameter
L = Length of Straight Side

P = (π × D) + (2 × L)

Note: To use this formula, first perform the multiplication inside the brackets, then add the two results together.

AREA OF A SEMI-CIRCULAR SIDED SHAPE

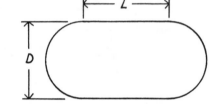

A = Area
D = Diameter
L = Length of Straight Side

$$A = (\frac{\pi \times D \times D}{4}) + (L \times D)$$

Note: To use this formula, first calculate each of the parts in brackets, then add the two results together.

VOLUME OF A CUBE

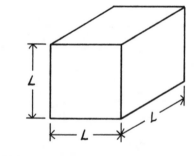

V = Volume
L = Length of a Side

V = L × L × L

VOLUME OF A RECTANGULAR SOLID

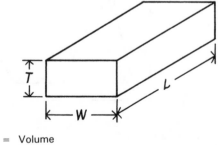

V = Volume
T = Thickness
W = Width
L = Length

V = T × W × L

Useful Information

VOLUME OF A SOLID PARALLELOGRAM

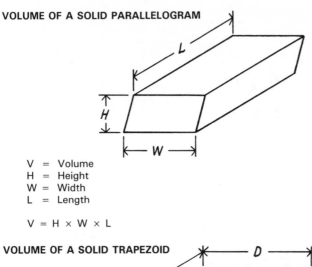

V = Volume
H = Height
W = Width
L = Length

$$V = H \times W \times L$$

VOLUME OF A SOLID TRAPEZOID

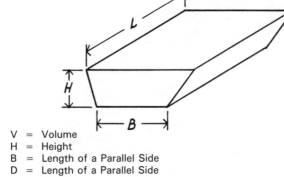

V = Volume
H = Height
B = Length of a Parallel Side
D = Length of a Parallel Side

$$V = 1/2 \times (B + D) \times H \times L$$

Note: To use this formula, first add the two figures inside the brackets, then perform the multiplications.

VOLUME OF A SOLID TRIANGLE

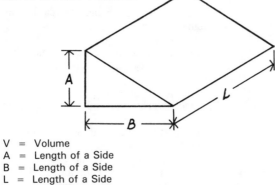

V = Volume
A = Length of a Side
B = Length of a Side
L = Length of a Side

$$V = 1/2 \times A \times B \times L$$

VOLUME OF A CYLINDER

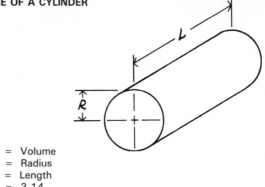

V = Volume
R = Radius
L = Length
π = 3.14

$$V = \pi \times R \times R \times L$$

VOLUME OF A SOLID HALF CIRCLE

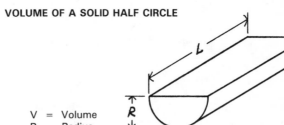

V = Volume
R = Radius
L = Length
π = 3.14

$$V = \frac{\pi \times R \times R \times L}{2}$$

Note: To use this formula, first perform the multiplication, then divide the result by 2.

VOLUME OF A SOLID SEMI-CIRCULAR SIDED SHAPE

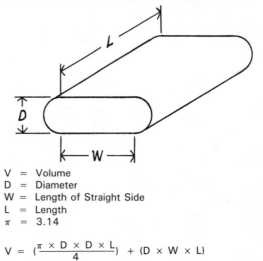

V = Volume
D = Diameter
W = Length of Straight Side
L = Length
π = 3.14

$$V = (\frac{\pi \times D \times D \times L}{4}) + (D \times W \times L)$$

Note: To use this formula, first perform the calculations inside the brackets, then add the two results together.

GLOSSARY OF TERMS

ACETYLENE GAS: A gas which is used in combination with oxygen as a fuel for welding and cutting.

ALLOY: An alloy is a blend of two or more elements, usually metal. Alloys are made to achieve certain special characteristics which do not occur in pure metals (see Stainless Steel).

ALUMINUM FLUX: A flux designed for brazing aluminum (see Flux).

ANGLE: The opening between two intersecting lines.

ANGULAR MEASURE: Refers to measuring the angle formed by two intersecting lines.

APPROXIMATE NUMBERS: Numbers that are not perfectly accurate. For example, the number 16,297 can be written as the approximate number 16 300 (see Rounding).

ARABIC NUMBER SYSTEM: The number system developed by the Arabs and now in common usage; made up of the ten digits 0, 1, 2, 3, 4, 5, 6, 7, 8, 9, (see Decimal Number System).

AREA: The entire surface measure of a shape.

ASTM: American Society for Testing of Materials. They perform a variety of testing services and set standards for industry. For example, in welding they have set up certain requirements and established various classifications for welding electrodes.

BASIC OPERATIONS: The mathematic operations of addition, subtraction, multiplication, and division.

BRASS: An alloy made up of copper and zinc.

C SHAPE: An abbreviation for Standard Channel. A structural steel product with a "⊏" cross section.

CAST IRON: Cast iron is an alloy of iron and carbon. To be classified as cast iron, the carbon content must be more than 1.7% (see Steel).

CENTIMETER: A metric measurement of length. It is 10 millimeters long.

CIRCULAR MEASURE: Refers to measuring curved lines.

CIRCUMFERENCE: The perimeter of a circle.

COLD GALVANIZING COMPOUND: This is a paste product which can be applied to steel for the purpose of protecting it from corrosion (see Galvanizing).

COLD ROLLED STEEL: This refers to a process used in making steel. Cold rolled steel is produced in its final form by passing it between successive rollers. Both the steel and the rollers are cold and this results in a very accurate steel product with a smooth, shiny surface (compare with Hot Rolled Steel).

COMMON DENOMINATOR: When two or more fractions have the same denominator, they are said to have "common denominators." For example, 3/15 and 2/15.

COMMON FRACTION: A fractional part expressed with a numerator and a denominator (compare with Decimal Fraction).

COMPLEX FRACTION: A fraction which has a fraction for the numerator and a fraction for the denominator.

CUSTOMARY SYSTEM: Refers to the common measuring system using feet and inches.

DECIMAL FRACTION: A number which is less than a whole number and is expressed with decimal. For example, 0.5 (compare with Common Fraction).

DECIMAL NUMBER SYSTEM: The number system in common usage having the ten digits 0, 1, 2, 3, 4, 5, 6, 7, 8, 9 (see Arabic Number System).

DECIMAL POINT: The point (.) in a decimal number which divides the whole part of the number from the fractional part of the number. For example, 12.75.

DEGREE: A unit of measure used for angles; 1/360 of a circle.

DENOMINATE NUMBERS: Numbers that represent measurements. For example, 17 feet is a denominate number; but the number 17 by itself is not a denominate number.

DENOMINATOR: The bottom number in a common fraction.

DIAMETER: The straight line distance across a circle and passing through its center.

DIFFERENCE: The answer that results when one number is subtracted from another (see Remainder).

DIVIDEND: In a division operation, it is the number being divided.

DIVISOR: In a division operation, it is the number that is doing the dividing. The dividend is divided by the divisor.

DOUBLE EXTRA STRONG PIPE: Pipe is generally classified as Standard, Extra Strong, and Double Extra Strong. The main difference being the wall thickness. For example, a Standard 2 in. pipe has a wall thickness of 0.154 in. while a 2 in. Double Extra Strong pipe has a wall thickness of 0.436 in.

EQUILATERAL TRIANGLE: A triangle in which all sides are the same length and all angles are equal (60°).

EQUIVALENT FRACTION: Fractions which are equal in value to each other. For example, 1/2, 2/4, and 3/6 are equivalent fractions (see Higher Terms).

EXTRA STRONG PIPE: Pipe is generally classified as Standard, Extra Strong, and Double Extra Strong, the main difference being the wall thickness. For example, a Standard 2 in. pipe has a wall thickness of 0.154 in. while a 2 in. Extra Strong Pipe has a wall thickness of 0.218 in.

FABRICATE: To construct by putting parts together.

FILLER MATERIAL: The rod which the welder adds to the weld in order to increase the amount of material making up the joint. Also, commonly referred to as a welding rod.

FILLET WELD: A weld with a triangular cross section.

FLUX: Flux is a material which is used to prevent the formation of oxides during the brazing (or soldering) process. It is usually in a paste form and is rubbed on the filler rod and the joint. Its main ingredients are boric acids, chlorides, fluorides, and wetting agents.

GA: Abbreviation of the word "gage" (see Gage).

GAGE: One of any numbering systems used to identify the various thicknesses of sheet steel.

GALLON: A standard American measurement of liquid capacity equal to 231 cubic inches. The Canadian gallon contains 277.42 cubic inches.

GALVANIZING: This is the process of applying a thin coating of zinc to steel or iron in order to protect it from corrosion. There are a number of methods for doing this, the most common being the hot dip method where the steel is dipped in molten zinc (see Cold Galvanizing Compound).

GEAR PUMP: A pump for delivering fluids. It causes a flow by passing the fluid between the teeth of two rapidly turning meshed gears.

GRAM: A metric measurement of weight; 1/1 000 kg.

HEAT EXCHANGER: A device used to transfer heat from a medium (such as water or air) flowing on one side of a barrier to a medium on the other side of the barrier.

HIGHER TERMS: A common fraction expressed as an equivalent fraction having higher values for numerator and denominator (see Equivalent Fraction).

HOT ROLLED STEEL: This refers to a manufacturing process used in making this steel. While the steel is still quite hot from manufacturing, it is passed through rollers to the required shape. Because it is being rolled while hot, this steel is rather dark, and is not as accurately formed as cold rolled steel.

HYPOTENUSE: The sloping line of a right triangle. It is the longest of the three sides.

IMPROPER FRACTION: A common fraction with a numerator larger than the denominator. For example, 7/6.

INSIDE CORNER: The side of a permanently bent piece of metal which is compressed and forms a concave shape.

INVERT THE DIVISOR: The process in the division of fractions whereby the divisor is turned upside down thus converting the division to a multiplication.

ISOSCELES TRIANGLE: A triangle in which two of the three sides are of equal length and two of the three angles are equal.

L.C.D.: Abbreviation for lowest common denominator (see Lowest Common Denominator).

L SHAPE: An abbreviation for angle. A structural steel shape with a "L" cross section.

LINEAR MEASURE: Refers to measuring the straight line distance between two points.

LOCK WASHER: A washer designed in such a way that it tends to enhance the holding power of a nut and bolt.

LOWER TERMS: A common fraction expressed as an equivalent fraction having lower values for numerator and denominator (see Equivalent Fraction).

LOWEST COMMON DENOMINATOR: A denominator for two or more fractions which is common to all the fractions and which is the lowest number possible.

MANUAL OF STEEL CONSTRUCTION: A book produced by the American Institute of Steel Construction Inc., providing a full description of structural shapes. The Canadian equivalent is produced by the Canadian Institute of Steel Construction.

MASS: A measure of the amount of material in an object.

METER: A metric measurement of length. It is equal to 1 000 mm (or 100 centimeters). It is about 3 in. longer than a yard.

METRIC SYSTEM: A simple and accurate system of measuring all things that are measurable. Also known as Si Metrics, or SI (which is an abbreviation of Systeme Internationale).

MIG: A common welding process also known as Gas Metal Arc Welding (GMAW). It uses a continuous consumable wire electrode while fed through the torch. The weld itself is covered and protected by a shield of gas which is also fed through the torch.

MIG WELDER: This refers to the torch, wire, feed, power, gas flow, etc., that make up the equipment necessary for MIG welding. The equipment can be completely automatic or semi-automatic.

MILLIMETER: The smallest metric measurement of length. It is 1/1 000 of a meter.

MINUEND: In a subtraction operation, it is the largest of the two numbers. The subtrahend is subtracted from the minuend (see also Subtrahend).

MINUTE: A unit of measure used for angles. 1/60 of a degree (see Degree).

MIXED NUMBER: A number consisting of a whole number and a fraction. For example, 5 3/8.

MULTIPLICAND: In a multiplication operation, it is the number that is being multiplied. The multiplicand is multiplied by the multiplier.

MULTIPLIER: In a multiplication operation, it is the number doing the multiplying.

NUMERATOR: The top number in a common fraction.

OF: A word used to express the multiplication of fractions. For example, 1/5 of 3/4.

OUTSIDE CORNER: The side of a permanently bent piece of metal which is stretched and forms a convex shape.

OXYACETYLENE: A combination of oxygen and acetylene gas which, because it burns at an intense heat, is used as a fuel in welding and cutting (see Acetylene Gas).

PARALLELOGRAM: A shape having four sides. The sides opposite each other are the same length and are parallel. The angles opposite each other are equal. Looks like a rectangle that has been tilted.

PAYLOAD: The maximum weight that a carrier (railroad car, airplane, etc.) can transport.

PERCENT: Parts of the whole of anything when it is divided into 100 equal parts.

PERIMETER: The distance around a shape.

PI: A Greek letter written as π and pronounced as "pie." It represents the numbers 3.14159 . . .

PLACE VALUE: The value of a number according to its place in a line of numbers. For example, the number 497 has three digits, each with the place value of hundreds, tens, and ones, respectively.

PROPER FRACTION: A common fraction with a numerator smaller than the denominator. For example, 2/3.

PROTRACTOR: An instrument used to measure angles.

QUOTIENT: It is the answer arrived at in a division operation.

RADIUS: The straight line distance from the center to the edge of a circle.

RECTANGLE: A shape having four sides. Similar to a square except that two sides are equally longer or shorter (see Square).

REDUCING: The process of changing improper fractions to mixed or whole numbers. Also, the process of expressing a common fraction in lower terms.

REMAINDER: The answer that results when one number is subtracted from another (see also Difference). Also, the amount remaining when a divisor will not divide evenly into a dividend.

RIGHT TRIANGLE: A triangle in which one angle is 90°.

ROOT OPENING: This refers to the gap that may (or may not) exist between two pieces to be welded. If the two pieces touch each other, the root opening is zero.

ROUNDING: The process of writing accurate numbers as approximate numbers. For example, the number 115,389 can be rounded to the approximate number 115,400.

ROUNDING THE DECIMAL: The process of writing accurate or unending decimal numbers as approximate decimal numbers. For example, 15.69032 can be rounded to 15.69.

S SHAPE: An abbreviation for Standard Shape. A structural steel shape with a slope on the inside faces of the flanges of 1:6.

SCREW MACHINE: A high-production, automatic machine originally designed to produce screws and other threaded fasteners. However, the name "screw machine" is highly misleading since they are used to form a large variety of parts including items such as welding tips.

SECOND: A unit of measure used for angles. 1/60 of a minute (see Minute).

SI: The metric system started about 200 years ago in France. As time passed, new versions and variations were randomly added to the system, causing unnecessary complications. Something had to be done, so in 1960, after lengthly international discussions, the International System of Units was established. The system is referred to as SI. It is the official, modernized metric system that is now replacing all former systems of measurement, including former versions of the metric system.

SILO: A tall, cylindrical structure in which grain is stored.

SOLDER: This is the filler metal used in soldering (which is a form of brazing). The solder melts below 800°F and joins the parts together. The solder is an alloy or combination of tin and lead.

SQUARE: A shape having four equal length sides. The opposite sides are parallel, all sides are the same length, and all angles in the square are 90°.

STAINLESS STEEL: An alloy steel which has a strong resistance to corrosion because of the addition of chromium as an alloy (see Alloy).

STEEL: Steel is an alloy of iron and carbon. To be classified as steel, the carbon content must be 1.7% or less. If it is more than 1.7%, the metal is classified as cast iron. Other ingredients may be added to the steel for special purposes (see Stainless Steel). Steel that contains no other ingredients besides iron and carbon is called plain carbon steel.

SUBTRAHEND: In a subtraction operation, it is the number being subtracted. The subtrahend is subtracted from the minuend (see also Minuend).

SUPPLEMENTARY UNITS: The radian and steradian are the two supplementary forms of measurement which are not included in the seven basic units of SI. They are used for measuring angles in two dimensions and three dimensions.

SURFACE GRINDER: A precision grinding machine on a pedestal. The workpiece is clamped in place on a movable work table and then passed under a revolving grinding wheel.

SURFACING: A welding process whereby weld material is deposited on a surface in order to build it up. For example, the worn surface of a bulldozer blade may be built up by surfacing and then machined to proper dimension.

TERMS: The numerator and denominator together are the terms of a fraction.

TIME CARD: A record of the hours an employee works on a day-to-day basis.

TOLERANCE: A specific range given of a dimension that a tradesperson may work within and still produce the desired results.

TRAPEZOID: A shape having four sides. Two of the sides are parallel, but the other two sides are not parallel to each other.

TRIANGLE: A shape having three straight sides.

U.S. STANDARD GAGE: One of the most popular gages used in classifying sheet steel thicknesses.

V-GROOVE JOINT: A type of joint where two flat pieces are beveled on their sides to form a V-shape.

VICE VERSA: The order of something that has been changed from that of a previous statement. For example, A distrusts B and vice versa; B distrusts A.

VOLUME: The amount of space an object occupies.

W SHAPE: An abbreviation for Wide Flange Shape. A structural steel shape characterized by a constant thickness of the flange. In general, it has a greater flange width and a relatively thinner web than S Shapes.

WEIGHT: The gravitational force exerted by the earth (or another celestial body) on an object.

WELDMENT: Any metal object made primarily by welding.

Odd Numbered Answers

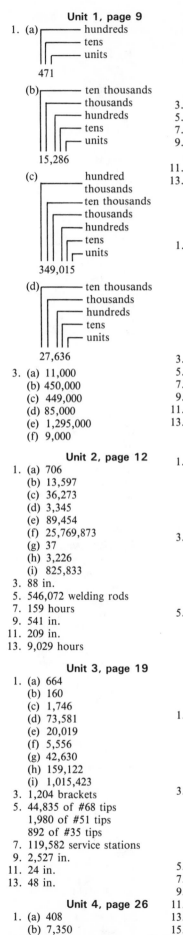

Unit 1, page 9
1. (a) hundreds, tens, units
   471
   (b) ten thousands, thousands, hundreds, tens, units
   15,286
   (c) hundred thousands, ten thousands, thousands, hundreds, tens, units
   349,015
   (d) ten thousands, thousands, hundreds, tens, units
   27,636
3. (a) 11,000
   (b) 450,000
   (c) 449,000
   (d) 85,000
   (e) 1,295,000
   (f) 9,000

Unit 2, page 12
1. (a) 706
   (b) 13,597
   (c) 36,273
   (d) 3,345
   (e) 89,454
   (f) 25,769,873
   (g) 37
   (h) 3,226
   (i) 825,833
3. 88 in.
5. 546,072 welding rods
7. 159 hours
9. 541 in.
11. 209 in.
13. 9,029 hours

Unit 3, page 19
1. (a) 664
   (b) 160
   (c) 1,746
   (d) 73,581
   (e) 20,019
   (f) 5,556
   (g) 42,630
   (h) 159,122
   (i) 1,015,423
3. 1,204 brackets
5. 44,835 of #68 tips
   1,980 of #51 tips
   892 of #35 tips
7. 119,582 service stations
9. 2,527 in.
11. 24 in.
13. 48 in.

Unit 4, page 26
1. (a) 408
   (b) 7,350
   (c) 3,920
   (d) 278,757
   (e) 174,915
   (f) 547,800
   (g) 4,748,576
   (h) 2,238,072
   (i) 1,876,070
3. 23,422 spot welds
5. 1,053 studs
7. 52,154 holes drilled
9. (a) 49 in.
   (b) 42 lbs.
11. 26,588 in.
13. (a) 765 lbs.
    (b) 2,852 lbs.
    (c) 3,500 lbs.

Unit 5, page 36
1. (a) 27
   (b) 12
   (c) 103
   (d) 1,150
   (e) 9,007   R-1
   (f) 68   R-30
   (g) 7
   (h) 123
   (i) 400
3. 45 lbs.
5. 16 pieces
7. 1,008 cans of flux
9. 18 in.
11. 5 in.
13. 48 lbs.

Unit 6, page 46
1. (a) 10/16
   (b) 100/200
   (c) 98/98
   (d) 5/16
   (e) 7/8
   (f) 1/2
3. (a) 4 1/2
   (b) 10 1/11
   (c) 17 5/8
   (d) 123
   (e) 3 1/2
   (f) 3
5. (a) 1/2
   (b) 5/6
   (c) 7/8
   (d) 1/9
   (e) 4/41
   (f) 1/128

Unit 7, page 51
1. (a) 11/18
   (b) 71/96
   (c) 168/219
   (d) 13/15
   (e) 2 23/50
   (f) 2 1/7
3. (a) 23 7/16
   (b) 87/112
   (c) 158 97/204
   (d) 49 13/16
   (e) 53/238
   (f) 2 417/2057
5. 89 13/32 in.
7. 57 17/64 in.
9. 235 21/32 lbs.
11. 27 9/32 in.
13. 52 19/32 in.
15. 505 55/64 in.

Unit 8, page 58
1. (a) 1/3
   (b) 1/2
   (c) 31/101
   (d) 5/8
   (e) 1/16
   (f) 10 69/340
3. (a) 2/3
   (b) 10 11/16
   (c) 7/15
   (d) 1/2
   (e) 12 2/3
   (f) 8 7/12
5. 7 37/64 in. and 4 7/16 in.
7. 3 1/2 games
9. 43 29/32 in.
11. 7 9/16 in.
13. Dimension A 19 13/32 in.
    Dimension B 16 9/32 in.
15. 22 39/64 in.

Unit 9, page 66
1. (a) 1/15
   (b) 7/20
   (c) 60/671
   (d) 1/4
   (e) 13/72
   (f) 1/60
3. (a) 20/63
   (b) 18 23/24
   (c) 25 11/64
   (d) 17/264
   (e) 2 5/8
   (f) 22 6/25
5. 1,750 lbs.
7. 72 1/4 hours
9. (a) 140 3/5 lbs.
   (b) 1,148 9/16 lbs.
   (c) 259 257/960 lbs.
11. 3,319 1/32 lbs.
13. (a) 2,243 7/15 minutes
    (b) 105,252 in.
15. 75 53/240 lbs.

Unit 10, page 74
1. (a) 1 1/15
   (b) 8/9
   (c) 20/21
   (d) 1
   (e) 9/16
   (f) 2
3. (a) 29/36
   (b) 1
   (c) 1 9/112
   (d) 1 3/5
   (e) 5/12
   (f) 11193/29356 =
       273/716
5. 17 7/12 in.
7. 49 pieces
9. 369 2/5 cu. ft.
11. 17 5/8 ft.
13. 10 pieces
15. 13 19/64 in.

Unit 11, page 83
1. (a) 100.01
   (b) 0.95
   (c) 14.00125
   (d) 3,219.125
   (e) 0.707
   (f) 0.1917
   (g) 0.866
   (h) 5.51563
3. (a) 0.5
   (b) 10.1
   (c) 0.03125
   (d) 0.703125
   (e) 0.875
   (f) 0.003
5. (a) 0.67
   (b) 0.25
   (c) 0.11
   (d) 0.99
   (e) 0.100
   (f) 0.49
   (g) 0.938
   (h) 0.429
   (i) 0.063
   (j) 0.488
   (k) 0.767
   (l) 0.752
   (m) 0.0156
   (n) 0.7333
   (o) 0.1235
   (p) 0.1111
   (q) 0.9659
   (r) 0.9688

Unit 12, page 86
1. (a) 11.1
   (b) 6.93
   (c) 354.388
3. (a) 222.49
   (b) 35.6231
   (c) 104.3088
5. (a) 6.787
   (b) 4.093
   (c) 0.559
7. 3.869 in.
9. $16.85
11. (a) $658.44
    (b) Greater by $158.44
13. 15.142 in.
15. 1.09375 in.
17. 15.8724 in.

Unit 13, page 92
1. (a) 0.4312
   (b) 3.69736
   (c) 70.356
   (d) 0.056042
   (e) 1,001.01
   (f) 218.75
3. (a) 1.3199
   (b) 11.857718
   (c) 195,000
   (d) 0.15
   (e) 65.20293
   (f) 10,010
5. 13.72 lbs.
7. $1,556.41
9. 23,800 cu. ft.
11. 942.081 lbs.
13. 2,550.95 lbs.
15. 1872.54846 lbs.

Unit 14, page 100
1. (a) 3,500
   (b) 390
   (c) 0.197
   (d) 1,000.1
   (e) 8
   (f) 0.0068752
3. (a) 3.000
   (b) 4.75

(c) 0.017
(d) 7.188
(e) 0.719
(f) 20.092
5. 312 cans
7. 27 passes
9. 623 items
11. 364 squares
13. 54 applications
15. 5.6233 in.

Unit 15, page 110

1. (a) 108 mm (4 1/4 in.)
   (b) 6 mm (1/4 in.)
   (c) 74 mm (2 15/16 in.)
   (d) 133 mm (5 1/4 in.)
   (e) 18 mm (23/32 in.)
   (f) 69 mm (2 3/4 in.)
3. (a) 423 in.
   (b) 1,476 in.
   (c) 544.5 in.
   (d) 206 14/23 in.
   (e) 6,042 in.
   (f) 49 1/2 in.
   (g) 63,360 in.
   (h) 781.2 in.
   (i) 3 3/16 in.
   (j) 353.64 in.
   (k) 145 in.
   (l) 5 5/8 in.
5. (a) 304.8 mm
   (b) 914.4 mm
   (c) 254 mm
   (d) 25.4 mm
   (e) 0.4 mm
   (f) 1 752.6 mm
   (g) 15.9 mm
   (h) 25 400 mm
   (i) 1 605.0 mm
   (j) 1 498.6 mm
   (k) 209.6 mm
   (l) 12 192 mm
7. (a) 15 3/4 in.
   (b) 1 in.
   (c) 48 in.
   (d) 80 1/8 in.
   (e) 19 3/8 in.
   (f) 7/8 in.
   (g) 19 14/16 in.
   (h) 32 1/16 in.
   (i) 1 in.
   (j) 855 8/16 in.
   (k) 21 1/16 in.
   (l) 1,010 11/16 in.
   (m) 11 21/32 in.
   (n) 58 1/32 in.
   (o) 730 6/32 in.
   (p) 9/32 in.
   (q) 31/32 in.
   (r) 3602 25/32 in.
   (s) 68 2/64 in.
   (t) 58/64 in.
   (u) 12 35/64 in.
   (v) 37/64 in.
   (w) 632 26/64 in.
   (x) 7802 49/64 in.

9.

Given Dimension	Conversion to Fractions	Maximum	Minimum
3.9 in.	3 90/100 in.	3 93/100 in.	3 87/100 in.
37.62 in.	37 62/100 in.	37 65/100 in.	37 59/100 in.
87.6 in.	87 60/100 in.	87 63/100 in.	87 57/100 in.
101.29 in.	101 29/100 in.	101 32/100 in.	101 26/100 in.

11. (a)
    7/16 in.
    29/64 in.
    1/2 in.
    33/64 in.
    17/32 in.
    9/16 in.
    4 31/32 in.
    17 3/4 in.

(b)
    0.438 in.
    0.453 in.
    0.500 in.
    0.516 in.
    0.531 in.
    0.563 in.
    4.969 in.
    17.750 in.

(c)
    11.13 mm
    11.51 mm
    12.70 mm
    13.11 mm
    13.49 mm
    14.30 mm
    126.21 mm
    450.85 mm

Unit 16, page 120

1. A 90°
   B 132°
   C 20°
   D 112°
   E 215°
   F 63°
3. (a) 28° 2' 20''
   (b) 3° 3'
   (c) 33° 59' 2''
   (d) 135° 2' 50''
   (e) 293° 43' 42''
   (f) 50' 32''
   (g) 11° 52' 28''
   (h) 86° 12' 58''
5. (a) 31° 14' 9''
   (b) 90° 36' 55''
   (c) 30°
   (d) 27° 45'
   (e) 51° 25' 43''
   (f) 45°
   (g) 8° 27' 59''
   (h) 42' 47'
7. (a)
   (b)
   (c)
   (d)
   (e)
   (f)
   (g)
   (h)
9. 22° 30'

Unit 17, page 131

1. (a) 5,328 sq. in.
   (b) 16,632 sq. in.
   (c) 1,400 sq. in.
   (d) 532 sq. in.

(e) 78 sq. in.
(f) 144,000 sq. in.
3. (a) 58 064 mm²
   (b) 422 451 mm²
   (c) 88 468 mm²
   (d) 650 321 mm²
   (e) 23 226 mm²
   (f) 10 000 mm²
5. 65,985 sq. in.
7. 68,888 sq. ft.
   9,919,872 sq. in.
9. Perimeter 29 564 mm
   Area 39 354 720 mm²
11. 13 222 mm²
13. Perimeter 347 1/2 in.
    Area 4,774 1/2 sq. in.

Unit 18, page 138

1. Perimeter 12 000 mm
   Area 6 000 000 mm²
3. Perimeter 336 in.
   Area 5,432 sq. in.
5. 55 ft.
7. Perimeter 256.05 ft.
   Area 2,812.5 sq. ft.
9. (a) Length 556.6 ft.
   (b) Area 5750 sq. ft.

Unit 19, page 148

1. 2,582 sq. in.
3. 76 1/2 in.
5. Area before holes 2,499 sq. in.
   Area after holes 2,016 sq. in.
7. Area in millimeters
   576 334 mm²
   Area in inches 893.3 sq. in.
9. Curved pieces 43 960 mm
   Straight pieces 80 000 mm

Unit 20, page 159

1. (a) 1,039 1/2 cu. in.
   (b) 1.0 cu. ft.
   (c) 13,464 gallons
   (d) 18 435 mm³
   (e) 119 gallons
   (f) 1.4 cu. ft.
   (g) 75.8 liters
   (h) 13,824 cu. ft.
   (i) 66,845 cu. ft.
   (j) 737 415 mm³
   (k) 4.6 cu. ft.
   (l) 269.28 gallons
   (m) 4.329 gallons
   (n) 155 cu. in.
   (o) 4,679.1 cu. ft.
3. 218 cu. in.
5. 1,416 cu. in.
7. 351 000 mm³
9. 23 minutes
11. 13,971 cu. ft.
13. 1,684 cu. ft.
15. 97 700 liters
17. 926 1/4 cu. in.
19. 1,955 gallons
21. 362 gallons

Unit 21, page 172

1. 15 kg.
3. 949 lbs.
5. 214.5 kg.
7. 370 lbs.
9. $101,691.65
11. 1,706.4 lbs.
13. 9,484 lbs.
15. 227 lbs.
17. 10 656 grams
19. 181.3 lbs.

Unit 22, page 189

1. 38 ft. 3 in.
3. 1 068 mm
5. 556 mm
7. 78 1/2 in.
9. 208 1/32 in.
11. 1 832 mm
13. 87 3/4 in.
15. 1 257 mm

Unit 23, page 199

1. (a) 3/25
   (b) 1/1000
   (c) 1/2000
   (d) 1/1600
   (e) 10 1/10
   (f) 1/140
3. (a) 25%
   (b) 1%
   (c) 3,333 1/3%
   (d) 256,050%
   (e) 57 17/19%
   (f) 9,906 1/4%
5. (a) 370.35
   (b) 47.66
   (c) 28.5
   (d) 84.33
   (e) 0.21012
   (f) 426.36
7. 189 parts per hour
9. 61%
11. $398,790,000

Unit 24, page 211

1. (a) Length
       Mass
       Time
       Electrical current
       Temperature
       Luminous intensity
       Substance
   (b) Length and mass
3. 1 000 mm
5. (a) 2 500 000 mm
   (b) 17.98 m
   (c) 0.012 8 km
   (d) 2 985.4 km
   (e) 10 000 mm
   (f) 0.059 5 km
7. (a) 39 24/64 in.
   (b) 15/64 in.
   (c) 4 32/64 in.
   (d) 3280 ft. 10 5/64 in.
   (e) 12 ft. 2 49/64 in.
   (f) 32808 ft. 4 51/64 in.
9. 1 000 ml
11. Gram
13. 77 590 000 mg
    (b) 39 250 g
    (c) 870 000 mg
    (d) 0.222 g
    (e) 0.001 kg
    (f) 0.044 kg